THE TRANSFORMATION OF THE WORLD

Science and Technology

SCIENCE AND TECHNOLOGY IN THE TRANSFORMATION OF THE WORLD

Edited by

Miroslav Pečujlić,
Gregory Blue and
Anouar Abdel-Malek

 ## THE UNITED NATIONS UNIVERSITY

in association with

St. Martin's Press   New York

© The United Nations University 1982
English language edition © Macmillan Press Ltd. 1982

All rights reserved. For information, write:
St. Martin's Press, Inc., 175 Fifth Avenue, New York, NY 10010
Printed in Great Britain
Published in the United Kingdom by The Macmillan Press Ltd.
First published in the United States of America in 1984

ISBN 0-312-70265-5

Library of Congress Cataloging in Publication Data

Main entry under title:

Science and technology in the transformation of the world.

   Based on the proceedings of a conference held in Belgrade and sponsored jointly by the United Nations University and the University of Belgrade.
   Includes indexes.
   1. Science—Social aspects—Congresses. 2. Technology—Social aspects—Congresses. I. Pečujlić, Miroslav. II. Blue, Gregory. III. Abdel Malek, Anouar. IV. United Nations University. V. Univerzitet u Beogradu.
Q175.4.S323   1984   303.4'834   83-40706
ISBN 0-312-70265-5

Contents

Preface	vii
The Transformation of the World: The Gearbox of Priorities Anouar Abdel-Malek	
Introduction Gregory Blue	xvi
Opening Addresses	1
I: Science and Technology as Formative Factors of Contemporary Civilisation: From Domination to Liberation	4

Preliminary Remarks
Keynote Address by Henri Lefebvre: Le Nécessaire et le Possible dans la Formation du Mondial
Rajko Tomović: Technology and Society
J. Leite Lopes: Science and the Making of Contemporary Culture
Y. Barel: Paradigmes Scientifiques et Autodétermination Humaine
A. N. Pandeya: Imagination, Insight and Understanding: Reflections on the Culture of Science in a Changing World
Discussion   31

II: Technology Generation and Transfer: Transformation Alternatives   39

Preliminary Remarks
Vladimir Štambuk: Conceptions of Scientific and Technological Development
Kenji Kawano: Science and Technology in Japanese History
Slobodan Ristić: Collective Self-reliance of Developing Countries in the Fields of Science and Technology
Vesna Besarović: Legal Aspects of the Transfer of Technology in Modern Society
Discussion   58

III: Biology, Medicine and the Future of Mankind   67

Preliminary Remarks
Bruno Ribes: La Maîtrise de la Vie, pour quoi Faire?

Yuji Mori: Restructuring a Framework for the Assessment of Science
and Technology as a Driving Power for Social Development: A Bio-
Sociological Approach
Viktor Milanović: Human Aspects of the Medical Sciences: Medical
Technology and the Physician's Responsibility
Discussion 81

IV: The Control of Space and Power 87

Preliminary Remarks
Osama El-Kholy: Towards a Clearer Definition of the Role of Science
and Technology in Transformation
José Silva Michelena: Science, Technology and Politics in a
Changing World
Zoran Vidaković: The Technology of Repression and Repressive
Technology: the Social Bearers and the Cultural Consequences
Luis Pinguelli Rosa: Nuclear Energy in Latin America: The
Brazilian Case
Discussion 111

V: From Intellectual Dependence to Creativity 120

Preliminary Remarks
Zvonimir Damjanović: Science and Technology as Organic
Parts of Contemporary Culture
Gregory Blue: Joseph Needham's Contribution to the History of
Science and Technology in China
Tetsuro Nakaoka: Imitation or Endogenous Creativity
Guillermo Bonfil Batalla: La Appropriación y la Recuperación
de las Ciencias Sociales en el Contexto de los Proyectos
Culturales Endogenos
Miroslav Pečujlić and Zoran Vidaković: On the Razor's Edge
Discussion 146

Appendix 1: Reports on Sections and General Report on the
   International Seminar 152
Appendix 2: Participants 163
Name Index 167
Subject Index 169

Preface: The Transformation of the World – the Gearbox of Priorities

The United Nations University's project on *Socio-cultural Development Alternatives in a Changing World (SCA)* was launched in mid-1978, within the framework of the Human and Social Development Program of the University, with a view to repositing the problematique of alternatives in human and social evolution as of the wide array of visions of our world, through its interwoven circles of civilisational moulds, geo-cultural areas, formations and nations.

Quite naturally, each of these interlinked circles comprises the socio-economic and political-ideological system as they obtain in the real concrete world of our times. However, the SCA Project is oriented towards a deeper level of analysis, deeper and more compelling too – i.e. the combination of the civilisational and geostrategic levels. Only by combining these two levels – the more obvious, traditional, level of the social sciences and the wider and deeper level of civilisation and geostrategy – does it appear possible to reach for the hidden part of the iceberg, for the very roots of the formative alternative schools of thought and action, deep at work. Only thus, and then, do we appear to be able therefore to mobilise their potentials towards a more humane, fraternal approach to the transformation of the world: to be precise, the non-antagonistic dialectical treatment of contradictions leading towards complementarity.

This First International Seminar of the series on *The Transformation of the World* deals with the dimension of *Science and Technology in the Transformation of the World*. It is thus the first of the series of six international seminars devoted to the implementation of the sub-project on *The Transformation of the World (TW)*, itself part of the UNU Project on *Socio-Cultural Development Alternatives in a Changing World (SCA)*, within the framework of the United Nations University's Human and Social Development Program, directed by Vice-Rector Dr Kinhide Mushakoji. A parallel series is devoted to the theme of the other sub-project *Endogenous Intellectual Creativity*, starting with the First Asian Regional Symposium, held at the University of Kyoto (13-17 November 1978), followed by the Latin American Regional Symposium at the Universidad Nacional Autonoma de Mexico (23-29 April 1979), the Arab Regional Symposium at Kuwait University (1981), etc. The series of five international seminars dealing with *The Transformation of the World* will comprise, after this first Seminar devoted to Science and Technology, the dimensions of: Economy and Society; Culture and Thought; Philosophy and Religion; the Making of the New International Order.

This First International Seminar was organised jointly by the United Nations University and the University of Belgrade, thanks to the perceptive help and deep commitment of Dr Miroslav Pečujlić, Rector of the University of Belgrade, our host and Chairman, and Dr Kinhide Mushakoji, Vice-Rector of the United Nations University for Human and Social Development Program.

In launching this series on the Sub-Project of *The Transformation of the World*, our SCA Project is aware that it thus fulfils an important part of the moral and scientific obligations of the international scientific community and of the United Nations University proper, at the very heart of our joint quest for a new international order, according to the fundamental decisions of the United Nations Organization and the Charter of the United Nations University, which coincide with the aspirations and decisions of the group of developing and non-aligned countries. The systematic, comparative and critical study of the different dimensions of *The Transformation of the World* is conceived of as the all-encompassing general frame and mould of the scientific and theoretical workshop now being developed, towards providing the international community with a deeper and more genuine understanding of linkages and differences, of our differing priorities, through their complex dialectical paths from contradictions to convergence. As such, this series of international seminars devoted to *The Transformation of the World* wishes to implement the aims and ideals of the United Nations University, as defined in its Charter:

> The University shall devote its work to research into the pressing global problems of human survival, development and welfare that are the concern of the United Nations and its agencies, with due attention to the social sciences and the humanities as well as natural sciences, pure and applied (Article I, point 2, UNU Charter).
>
> The research programme of the institutions of the University shall include, among other subjects, coexistence between peoples having different cultures, languages, and social systems; peaceful relations between States and the maintenance of peace and security; human rights; economic and social change and development; the environment and the proper use of resources; basic science and technology in the interests of development; and universal human values related to the improvement of the quality of life (Article I, point 3, UNU Charter).
>
> The central character of our times, of the real world in our times, resides in the transformation — not evolution or transition (all historical periods are periods

Preface

of transition) — of all dimensions of the life of human societies. To be sure, this transformation, acknowledged by all quarters and groups all over the world, is neither unilinear nor synchronic. At the first level, we are witnesses to major differences in the quality, quantity — and more so, the tempo and impact — of processes of transformation in different sectors of social life and activity: economic production; patterns of power; societal cohesiveness; cultural identity; civilisational projects; political ideologies; religions; philosophies; myths, etc. — in short, all sectors of what is usually termed the infrastructure and superstructure of society. At a second, more visible and forceful, level, we do acknowledge distinctions between different types of societies, e.g. in the different types of socio-economic formations and the accompanying political ideologies (basically capitalism, liberal capitalism and monopoly capitalism; and socialism, national progressive socialism and communism). And even more so, in the hitherto neglected dimension of civilisational, cultural and national specificity, we encounter major, more resilient and protracted, sets of differences.

This transformation of the world can be recognised in the following three sets of factors, which lend themselves to being recorded according to different conceptions of priorities:

(1) The resurgence of the three continents of Asia, Africa and Latin America to contemporaneity, both in the socio-political and civilisational-cultural fields. The historical processes of national liberation and independence, coupled with national and social revolutions, have gathered momentum since their inception in modern times, during the early part of the nineteenth century, until they became the dominant factor of contemporary history from the years following 1917 and especially in the period 1945-1973.

This vast transformation has been seen by Western specialists as a socio-political process within the traditional conception of the world's history (as consisting of one centre — Europe, later Europe and North America, i.e. the Western World — and its periphery, the Orient, i.e. Asia, Africa, the Arab-Islamic world, later joined by Latin America). The three continents were emerging but what was/is emerging is seen in socio-political terms.

On the other side of the river, especially in the Orient — Asia, Africa and the Arab-Islamic world — this process of emergence was seen essentially as a process of renaissance of either culture or civilisation, as in the Arab and Islamic *Nahdah* and Meiji Japan, in the Chinese Cultural Revolution and in the upsurge of Africanism, while Latin America's quest for identity has brought to light the hitherto hidden Indian and Indo-African elements of the culture.

(2) A parallel, major set of formative factors in this transformation appears to have developed between 1848 and 1973, and especially from October 1917, the date of the first socialist revolution in history. The hitherto equanimous front of the bourgeoisies in power was suddenly faced with the eruption of the labouring people into power, coupled with a populist *Weltanschauung* geared towards a persistently more humane life for the have-nots. Sixty years later, nearly half of mankind lives under socialism — four-fifths of whom belong to Asia and Africa.

(3) More recently, a third set of factors has become more visible, centring

upon the immense progress accomplished in the fields of science and technology. Here, again, while certain advanced Western countries opted for such denominations or descriptions as the 'scientific and technological revolution' or 'post-industrial society', on the other side of the river the vision remained paradoxically nearer to more realistic approaches, using the more traditional concepts of 'revolution', 'development', 'social transformation', within the implacable parameters of geopolitics. Yet none would deny the message and ever-growing influence of the application of modern technologies in our world, in the very fabric of our individual lives through the complexity of societal processes.

A long way, indeed, from the ethos and tonality dominant in 1945 — a long, long way.

Neither atomic clouds above the North Pacific, nor the hideous convulsions of traditional imperialism and colonialism in Asia and Africa, nor the liberation of the largest country in the world in 1949 could bring sense to the massive thrust in Western advanced industrial societies towards productivism, consumerism, hedonism. Finally, the Golden Age of Man-as-demiurge had been reached, the very frontiers of the Promethean concept so persistently at the heart of the Western civilisational project, from the age of maritime discoveries and the European Renaissance to Yalta. And the instruments of this historic fulfilment were none other, precisely, than science and technology as the driving forces in the second stage of the Industrial Revolution.

If Man was finally the master of Nature, the conqueror of the universe, geared to achieve all the panoplia of pleasures he could dream of, what, if any, would be the use in keeping such 'archaic' concepts and moulds as nation and State, the family, working people and the tools of exploitation, to say nothing of such 'distant' objective superstructures as philosophy, religion, the humane values of love and fraternity, equity and peace — let alone the civilisational project? In spite of the powerful waves making for the transformation of the world, few, or at best a large minority, were listening to the 'voices of silence', to Joseph Needham's favourite Confucian saying, 'Behave to every man as one receiving a great guest', to Chou En-Lai's 'Don't forget the well-digger when you drink water'. Or was it because of them?

Yet, in less than ten years, ethos and tonality have shifted decisively towards the rancourous penumbra of the 'Crisis'.

Which crisis?

In the North, leading epigones are busy mending fences. Oil, raw materials, the receding markets, non-competitive old industrial plants: such was the verdict, with some lonely exceptions. And this verdict was echoed by a large proportion of audible voices in the South, the good 'Westernised modernisers', busily engaged in reciprocating, even if now with more strident voices.

That the crisis could be that of Civilisation itself was here and there mentioned. But this Civilisation was conceived of as that of the still hegemonic 'centre', as against the underdeveloped or developing non-Western 'periphery', provoking a mixture of reluctant acceptance and anguished self-interrogation. That the crisis might be, perhaps, that of the civilisational project of the hegemonic West itself,

Preface xi

much more so than its actual hegemony and precedence, in power terms, began to emerge, here and there, followed by intense reactions of either apocalyptic pre-visions – if the Western civilisational project was in crisis, how on earth could mankind seek alternatives? – or derisive comparisons and strictures facing the incoherence and lagging behind of the non-Western world.

For it is true that major parts of the underdeveloped non-Western societies are still caught in the mirage of reductionism, busily imitating the advanced industrial societies of the West (as if history was indeed repetitive, its formative historical moulds and real concrete processes amenable to copying) plummetting towards limitless productivism, consumerism and hedonism, equating progress with profit, domination, the ghettos of individualism – the negative mind. As if nothing could be different from that very combination of factors eroding in depth self-assurance, popular and national self-reliance, the feeling of security, the hope for a more fraternal and equable future for the majority of mankind – the taming of the 'acquisitive society'.

Hence the quest for alternatives.

In the field of science and technology, the quest is now towards 'alternative technologies' or 'appropriate technologies', with a sprinkle of 'radical technologies'. If a set of scientific applications, of technologies, is to be sought to escape the dilemnae of advanced Western industrialised societies, then they could only be – in the reductionist approach – an 'alternative' set of technologies, parallel to the advanced Western varieties. And this set could mercifully be located through the concept of 'appropriate' technologies. 'Appropriate' to what? 'Appropriate' to whom? 'Appropriate' for which purposes? 'Appropriate' according to which, and whose, criteria? To be sure, history has it that the great majority of the nations of the 'three continents' can hardly echo the procedures which enabled the West, in five centuries, through the concentration of historical surplus value, to develop gradually its modes of capital-intensive productivity. The humane uses of human resources, in the advanced nations of Asia, Africa and Latin America, as of the socio-economic restructuring of the societal fabric, is now seen to be more beneficial than was hitherto imagined in bridging the gaps between rationality and fraternity, in giving a more humane vision of social dialectics than was hitherto prevalent.

Yet the temptations, traditions and fringe benefits of survival imitation, the reluctance to use vision as a tool for our futures, remain immense. For, then, the question would be: 'To which technology does vision belong?'

The growing criticism of the impact of science and technology on modern societies and human life, through its diversity and different motivations, gives an impression of leading towards a growing ambiguity. For, although this impact, through hegemony, has had its negative and pernicious effects in the underdeveloped areas, in the 'three continents' of Asia, Africa and Latin America, to this day, whether through the direct domination of imperial powers or their more systematic pillaging by multinationals, the more recent mounting criticism has come from the developed areas, from the core of the West.

The tonality here is of alarm, and the contents ethical and normative. Industrialisation and urbanisation have led to ecologism; atomic armaments and nuclear energy, to the quest for pacifism; consumerism and individualism, at the time of the energy crisis, to the pursuit of more humane, low-key participatory patterns of social interaction. And it is from the core of the more advanced industrialised societies of the West that the most ruthless indictments of science and technology are nowadays being launched.

On the other side of the river, in Asia, Africa and Latin America, the mounting wave of national movements, often coupled with social transformation or revolution, has always clearly proclaimed its desire — in all countries nations and societies in the so-called 'South' — to achieve contemporaneity, to modernise, as of the paths and potentials of its variegated national-cultural specificities grounded in the depths of the historical field. And the instruments and means to achieve this global legitimate desire have been defined simultaneously, in the inner circle, as the creation and reinforcement, or revival, of a stable centre of national social power, the independent national State of the tricontinental area in our times, to be accompanied in the outer circle by careful prospecting of the realities of the balance of power, of the evolving patterns of dialectical inter-relations between major centres of power and influence in our times.

For here, more than ever before, more than anywhere else, more than in any other field at any other time in the history of mankind, the massive unanimous protracted consensus of Asia, Africa and Latin America, of the group of developing and non-aligned countries, lies in the coupling of independent national power of decision, only feasible on the basis of an advanced level of science and technology in the fields of economic production, State structuring and mass onslaught on illiteracy and backwardness, with a meaningful and equable share in policy-making at world level. Such are the roots, visible for all to see, of U Thant's call for what was then labelled the 'New International Economic Order', what has gradually become the 'New International Order', at the time of the transformation of the world. A close scrutiny of major decisions and their philosophy in the series of major conferences from Bandoeng to Belgrade, Colombo and Havana, of the socio-political contents of the politics put forth by all national independent States of these areas (four-fifths of mankind, through the deep diversity of their socio-economic and political-ideological regimes, with the exceptions — isolated societies, or compradors fringes) bears witness to this reality.

The call has been, and remains, for a realistic political approach of human society in our times, a deep desire to use fully the contributions of science and technology as a means to secure a wider and greater share in power of decision at world and regional levels — more often than not, attuned to civilisational visions, cultural traditions and national parameters — but never evasive about the deep structural integrated inter-relations between power and culture, at the heart of all problems of human and social development.

As a matter of course, both sectors of world societies — the so-called 'North', and the so-called 'South' — meet along the more general issues, such as nuclear disarmament, or the acknowledgement of the need for more rational relations

Preface *xiii*

between the two sectors. But, short of the extreme parameters of annihilation, the rise to contemporaneity of Asia, Africa and Latin America is seen by the formative endogenous schools of thought and action in these continents in terms altogether different from those of the dedicated minority groups in advanced industrialised societies, who are justly rebelling against the dangers inherent in their societies and its civilisational project. At the same time, the power structures of modern advanced industrialised societies, with the broad support of the wide masses of the population, including the working people industry, agriculture, and the services alike, are, in fact, persistently taking action to reach an ever-growing level of scientific and technological sophistication in all fields of social life, with a view to ensuring their continuous hegemony through the coming generations and, hopefully, centuries.

Here lies the principal contradiction between 'the two sides of the river', between the hegemonic power centres of advanced industrialised societies, on the one hand, and the rising national independent influence centres of the heretofore marginalised cultures and societies of the world; while the secondary contradiction seems to lie, at a much lesser degree of intensity and, perhaps, a higher level of ambiguity, between the humanistic minorities of advanced industrial societies, on the one hand, and the Tricontinental area, on the other hand.

Clearly this area of contradictions is of crucial importance in defining the problématique of our joint prospection. It is here, so we feel, that the confrontation of analyses, the uses of meaningful comparatism, the perceptive understanding of different types and scales of priorities, can be of genuine benefit for the international community, for deeper understanding of the transformation of the world in our times. It is here, so we feel, that the challenges and difficulties of the dialectics of tradition and modernity, specificity and universality, are calling upon us to search for the deepest roots, the hidden part of the iceberg, as it were.

This is a task of vital importance in our times: an imposing challenge to the international intellectual community: the duty of all concerned citizens towards their nations, peoples and cultures.

As Socrates, the master of interrogative dialectic, taught, many a century ago, 'everyone acts according to his knowledge'. And we now know that Louis Aragon is right in asserting that 'the future has not already been lived'. If knowledge, philosophical knowledge of the inner workings of societies in our time, is indispensable and worthy of continuous attention, could it be stated in confidence that a better knowledge, a deeper understanding of the present, both as history and potential future, could chart the path towards more rational and humane endeavours?

To this task of paramount importance, the historic task of bridge building, our UNU Project *Socio-Cultural Development Alternatives in a Changing World* (SCA) is, above all, dedicated. For our's are the challenges and promise to construct jointly what we would propose to define as the gearbox of priorities: to bring together in meaningful interaction, towards complementarity, the widely different schools of thought and action in this, our world, rooted in civilisational, cultural, national specificities, socio-economic formations, political systems, philosophic, religious, ideological visions of the world, scientific, theoretical and methodological conceptions.

As we approach the practical aspects of our research, the more practical, policy-oriented aspects of our endeavours, we are bound to face the basic dialectic between specificity and universality under the guise of what we would propose to call the dialectics of priorities. It is obvious that policy definition, differences in standpoints at programmatic and practical levels alike, relate directly to, and are grounded in, what appears at first sight to be a difference in priorities. Then, how can we come to grips with this contradictory aspect of our problématique?

(1) The first level of analysis will deal with the definition of *categories* of priorities:

(a) Some would incline to put the first category in the domain of production, economics and their accompanying technological and scientific aspects. We would have, here, *inter alia*, productivism and consumerism; low-key development and hedonism; individual patterns of economic organisation; and collective and State patterns; etc.

(b) The political dimension proper vividly obtains inasmuch as priorities take their shape through political decisions by the concerned bodies and institutions of all societies. Usual distinctions between liberal and autocratic, democratic and dictatorial, populist and despotic, consensus and élitist, etc., are naturally obtained and are of direct relevance to the definition of priorities.

(c) A third category can be located in the realm of culture, thought, philosophy, ideology, religion, as of their formative historical moulds: this is where we find the greatest number of differences, echoing the differentiation of human societies in nations, cultural areas and the proliferating *Weltanschauungs* cutting across the different levels of this realm.

(2) We would then address ourselves to a second level of differentiation, i.e. the different *types* of priorities:

(a) A first general type in priorities is the static-conservative type, i.e. the type of priorities more concerned with the maintenance of societal cohesiveness, socio-economic and political-ideological systems, either facing the mounting wave of new transformational and radical demands, or just as an expression of the necessity to preserve achievements and acquisitions which had been the results of lengthy processes of transformation before the crystallisation into a viable new order. The different legitimisations of this conservative approach clearly mean that the contents of what is sought to be conserved can be, and are, profoundly different – yet appear for a certain time more static than their proclaimed aims and contents.

(b) A second general type in priorities is the radical type, oriented towards the transformation of societal moulds. Here priorities will often appear in parallel, dual, contradictory patterns, and not just as different stages in the same type of priorities, as is often the case in the conservative type of priorities.

Preface

(3) Enough has been said, even though sketchily at this stage, to give a sense of the immense complexity of defining priorities, let alone making sense of their differences. Yet the most disconcerting aspect in priorities appears to be the aspect/dimension of *tempi*. For while the difference in priorities — through their different categories and types — can be understood, and even accepted, as a rational discourse, the operational position of priorities through the *time-dimension*, i.e. *the transition from choices to action, from decision to praxis, represents the hour of truth in the dialectics of priorities.* And here, again, it is important to note that different *tempi* are not derived only from the subjective moment of decision making: they are rooted, objectively, in the objectivity of the geohistorical constraints defined in the outer and inner circles of social dialectics in different societies of our world, as well as the different visions obtaining within these societies of the alternatives ahead of them.

(4) Hence the quest for a mediation which combines the distinctions in a way that can make them understandable, acceptable to a reasonable extent, or at least properly perceived within their own objective legitimacies. The intent here is not to solve the dialectics of priorities but rather to clarify the hidden part of the iceberg which does forcefully make for contradictions, oppositions and frontal antagonisms. A central task of the SCA Project has therefore been seen as the gradual construction of the 'gearbox of priorities', a gearbox whose component parts are none other than the differentials representing the above-mentioned categories and dimensions of the dialectics of priorities.

As we set out to initiate the series of international seminars on *The Transformation of the World* with the study of the domain of science and technology, let us remember the hope and urgency, the reality of our real concrete world, the vision of our converging futures.

In fraternal amity and realistic lucidity, let us therefore join hands.

Paris and Cairo, January, 1982   Anouar Abdel-Malek,
                                                                  Project Coordinator

Introduction

Sponsored jointly by the United Nations University and the University of Belgrade, the conference recounted in these pages and devoted to examination of the theme of 'Science and Technology in the Transformation of the World' was the first of a series of six organised by Prof. Anouar Abdel-Malek around the sub-project on 'The Transformation of the World'. This first seminar of the series was held in Belgrade from 22 through 26 October 1979 and it was attended by some 70 participants, most of whom were academics. Of these participants, approximately half were guests from abroad – Japan, Western Europe, the Islamic world, India and Latin America all being well represented. The other half of the participants consisted of a friendly and extremely hospitable contingent of distinguished Yugoslavians, who were serious-minded but lively and endowed with that talent for languages for which their countrymen are so well known.

Apart from the formal opening ceremonies (which were held on the University's premises in the 'Old City'), all of the conference's sessions took place in the comfortable facilities of the Hotel Jugoslavija, which faces the Bulevar Lenjina in Belgrade's 'New City' (that section constructed since Liberation in 1944) and which from the rear overlooks the Danube close to where the latter is joined by the Sava. After the opening ceremonies on Monday morning, five plenary working sessions of four hours each were convened during the next two and a half days, ending on Wednesday evening. Thursday was devoted to informal workshops, each of which examined questions that had developed out of one of the working sessions held on the previous days. Friday was taken up with the drafting of reports on the work of the conference and with a final plenary session during which these reports were presented, discussed and accepted.

What follows is a detailed general report based on the (revised, but unedited) typescript of the *Proceedings* of the conference. This report was commissioned with the idea of providing a fairly readable account, of a reasonable length, of the way in which the various arguments were expounded and developed; and it is hoped that it will be useful in making the work of the conference available to the general reading public. The format of presentation adopted has, in general, been that of keeping the chronological sequence of the various sessions intact, although the order of the several position papers within each section has been altered for purposes of logical exposition. The account of each session of the conference is here preceded by a brief introduction which attempts, as it were,

Introduction xvii

to put into relief the most significant questions treated in that particular session. Following each introduction, the positions taken by the various participants are summarised, starting with the position papers and then proceeding to the discussion. The summaries here on the average amount to approximately one-quarter of the original length of the papers and transcribed interventions; direct quotations have been used wherever possible, and, in general, I have tried within each summary to adhere to the original in regard to both mode of expression and sequence of argumentation.

In editing any such large body of material, however, and especially one covering such a wide range of subjects, one inevitably leaves aside much that is fascinating, either because of requirements of space or balance or because of the necessarily limited scope of one's own vision. Therefore, the interested reader is advised of the forthcoming publication of the integral text of the conference *Proceedings*, which can be consulted on any question of particular interest; and, this having been mentioned, perhaps it can be suggested that the present text might usefully serve not only as a detailed account of the conference in its own right, but also as a companion volume to the full *Proceedings*.

The great differences between spoken and written language are of course well known; and while I have quoted many oral interventions directly, I have not hesitated to rework these, quite drastically at times, according to accepted norms of written English. This has been all the more necessary, and frequently for written as well as for oral pieces, in so far as most of the participants in the conference have had to communicate in a language other than their mother-tongue. On the other hand, several papers and speeches appear officially in the *Proceedings* of the conference in languages other than English; translations of passages from such pieces are my own and at times, I fear, inevitably lack the flavour and elegance of the originals.

In general this report is composed of a relatively high proportion of directly quoted material from the conference, but I have tried to keep the text as continuous and readable as possible. The use of ellipsis dots has therefore been kept to a minimum, except in cases in which material has for one reason or another been deleted from within a passage that is quoted. At the same time, there are many places in the report at which either a sentence or a paragraph ends with quotation marks while the following one begins immediately with new quotation marks. I'm afraid that in a text of this sort such oddities are rather difficult to avoid, and I can only hope that readers will have come across similar practices in journalistic writing and will be able to accommodate them here without much annoyance.

It should also be mentioned that, for one reason or another, several papers and interventions presented to the conference were not included in the typescript of the *Proceedings*; I have therefore been unable to give an account of those pieces in the present work. This has been the case with the position paper by Dr Imré Marton and with the written text of the paper by Dr A. N. Pandeya; it has likewise been the case for oral interventions by Drs Lefebvre, Barel, Marton and Besarović.

The reader should be aware from the outset that the conference's attention was focused primarily on the *social and political significance* of science and technology. In other words, if we were to define science as a process by which human societies tend to understand objectively and, hence, to gain control of Nature (and if we are allowed a very sweeping generalisation), then it can be said that the participants in the conference concerned themselves more with the activities and organisation of the subjects (i.e. human societies) than with the effects of such activities on that generic object, Nature, or with the material instruments for obtaining such effects. As noted in the discussion during the fifth session, most of the participants were in fact social scientists; and even those whose background lay in the natural sciences or engineering spoke during this gathering – and necessarily so – within the framework of the social sciences. Thus, one could say that, while micro-chips, nuclear energy and genetic engineering marked the contours of the landscape, the general terrain of the conference would have to be sited somewhere between what Yves Barel termed 'socio-epistemology' and what is generally known as 'science policy'. Unless I am mistaken, the choice of this terrain was a reflection of the fact that social movements striving for emancipation, and especially those in the Third World, are being forced to map theoretical and practical orientations which can serve to guide them in avoiding the various forms of scientific-technological determinism (both pessimistic and optimistic) coming more and more into relief now that economic determinism in its more general guises is being challenged increasingly, in the UN as elsewhere.

It is going to be obvious throughout this report that the politics of science and technology are inextricably linked to global politics in general. It would, however, be quite mistaken to expect the participants in the conference to have unanimously shared a common estimation of the present nature of this linkage or of the priority of tasks necessary for its progressive transformation. For example, proposed courses of action inevitably departed from particular analyses or readings of the present situation. Many participants spoke of political structures of domination and dependence as constituting the foundation of inequalities in the world today, while others put the emphasis on economic differentials or on disparities in scientific-technological potentials; still others stressed monopolistic control of informational resources, and some cited differences in levels of education as the root of the problem. While disparities at each of these instances are undoubtedly essential aspects of the structure of global inequality, the variety of specific weights assigned to each of them by different participants can be taken as a gauge of the contradictions within the assembly. Despite such differences, however, I believe that the conference functioned usefully as a forum not only for exchanging opinions, but also for mutual understanding and education.

I have not hesitated to express my own views and even to editorialise now and again in the course of this text; but I have tried, in general, to delineate them clearly from the views of others, either by restricting them to the intro-

Introduction

ductions to the various sections or by inserting them into the summaries as obvious commentaries on the original texts. Nevertheless, the nature of a report like this has made it inevitable that my own positions should surface spontaneously, and it is probably only fair to make them explicit very briefly here. I will confine myself to three quick indications regarding questions of quite different magnitudes. First of all, I have at various points in the report referred to a distinction between traditional sciences and modern science (as it has grown up since the seventeenth century). I consider this distinction to be of methodological as well as of historical significance, for it is vital to a correct understanding of the unity of scientific endeavour. Further clarifications are to be found in the summary of my own paper in the fifth session. Secondly, a capacity to generate scientific and technological knowledge and innovation is an essential facet of any healthy productive system today; and to prevent a society from attaining such a capacity is in effect one substantive way to keep it not only backward, but also weak. Moreover, mastery of contemporary science and technology requires a genuine, dedicated respect for their objective specificities, which cannot be argued away. On the other hand, science and technology exhibit both productive and destructive aspects; and each of these aspects is itself problematical, given the various striking social contradictions in the world today. Science and technology can thus contribute to production which is either useful or wasteful, and either beneficial or harmful to human well-being; they can likewise contribute to the destruction of those phenomena which constitute obstacles to human well-being and creativity, or they can contribute to the destruction of some of the most precious qualities of mankind – and, indeed, even to the attempted destruction of entire societies. Which of these ends science and technology are to serve depends primarily on which social forces control their use and development. While it may thus well be true that science and technology can and even must be integrated to serve the often contradictory aspirations of different social forces, it is nevertheless incorrect to interpret this as meaning that they are socially neutral in any absolute sense: they are, rather, immense sources of power for whoever controls them. And at our present moment in history the progressive potentials of science and technology can in my opinion only be realised if they are placed at the service of the mobilised forces of the working people, both on a national and on a global scale. Lastly, then, I believe that the peoples of the Third World today as a body constitute the *main force* for the progressive transformation of the world, and the primary positive vocation of science and technology must be realised precisely in meeting their needs. Despite the machinations of those with vested interests to protect, this mammoth force in the underdeveloped parts of the world is more and more being joined by progressive forces in the developed world as well. It is thus possible to speak of the emergence of a global united front; and both positive and negative demands in regard to scientific-technological activities necessarily comprise important planks in the platform of this front. In my opinion, the chief forces opposed to the forward movement of the front as a whole are the two superpowers; and which one poses the most immediate danger to a particular country at a given moment undoubtedly depends on the conjuncture of a number of concrete factors. In general, however, the oppression and exploitation of the dependent countries in the world is perpetuated not only by the various forms of collusion between the two superpowers, but also by their rivalry — since (in the recent words of a senior non-aligned head of state) whenever great powers struggle to expand their spheres of influence, small nations are victimised and their interests damaged. The peoples of the Third World are thus striving to stand on their own feet and to unite, in order to throw off all forms of foreign domination — including the monopolist control of science and technology — which prevent them from occupying their rightful place in the world. At the same time, all progressive forces in the world must face the imperative task of fighting a positive struggle to unite all anti-imperialist and anti-hegemonist forces for the task of preventing another world war.

As a final word before moving on to the body of the report, I should like to express my thanks to Ludgard De Decker for her generous help in all phases of the preparation of this text.

Cambridge, 1981 Gregory Blue

Opening Addresses

On the morning of Monday 26 October, the participants proceeded by coach into Belgrade's Old City to the Secretariat of the University for the formal opening ceremony of the symposium.

Mr Živorad Kovačević, the president of the City Assembly, began the proceedings by welcoming the participants to Belgrade in an address which succeeded in raising several major themes that would be discussed in the days to come. Perhaps the most important of these themes concerned the contradictory aspects which science and technology can exhibit. On the one hand, "science and technology ... are parts of a new civilisational wave, [and they] are becoming a driving force of development, a strong lever for the humanisation of the world, for the liquidation of poverty and hunger, and for reducing the gap between rich and poor countries — this volcanic contradiction of our epoch". On the other hand, however, it is a dangerous illusion to believe that 'technique by itself' can solve all of our problems. For in spite of its great potentials, "it can be abused; it can become destructive of nature," and it can be "transformed into a powerful instrument of domination... over people and whole communities; it can be used" — as indeed it is — "in favour of privileged groups and countries". One must face this dilemma squarely and realise that neither blind faith in science and technology nor renunciation of them will provide a lasting solution to the problems of the peoples of the world.

Technological growth will contribute to human progress only if it is linked to the real needs of ordinary people; and if this requirement is to be met in the world today, then it is essential "that every country develop its own creativity and not merely adopt foreign patterns of industrial urban development". Our times have of course become an epoch of technological, economic and cultural interdependence, and it will be most useful for nations to draw on universal scientific achievements. "But the world will become a real human community only provided that each country enriches it by its own authentic, unique creativity, looking for answers to problems that we all have to face...."

If universities can break out of their traditional isolation and attune themselves to social needs, they can play a leading role in discovering new possibilities of development by linking global achievements in science and technology with the need in their own societies for the development of all creative potentials. Mr Kovačević concluded his speech by citing the example of one such institution, namely Belgrade's Centre for Urban Technology, whose principal task is to implement theoretical knowledge accumulated around the world, so that the citizens of the city — producers and consumers — can more effectively participate in making the vital decisions that affect their everyday life.

Such participation in decision-making has, of course, not yet been won as a common right for the whole human race, and this fact goes far in explaining the fundamentally ambiguous role of science and technology in the world today. For example, it is not difficult to imagine why systematic efforts were made to prevent the citizens of Pennsylvania from learning about the risks attached to the nuclear power station at Three Mile Island. As noted by Dr Pavle Savić, chairman of the Serbian Academy of Sciences and Arts, science and technology must be seen in the light of the sharp social contradictions which characterise our times, both nationally and internationally. We inhabit a world in which the lives and the co-operatively productive labours of millions of people are controlled and directed by a few tightly closed groups which attempt to suppress all efforts at overcoming their strangle-hold of privilege and power; this is a world in which "superdeveloped and extremely underdeveloped human communities exist simultaneously", a world in which "millions of men, women and children are dying of starvation and disease", while "a minority enjoys the benefits of accumulated wealth". In such a situation, it is not surprising that the results of the development of science and technology are not quite 'universally beneficial': consider the stockpiles of nuclear weapons, consider napalm, etc. Taking up a topic that would be developed by Prof. Rosa during the fourth session, Dr Savić pointed out that the technology of nuclear energy remains a virtual "monopoly of the superpowers", and he likewise deplored the widespread environmental pollution brought about in developing countries by the use of obsolete technologies which in economic terms only serve to strengthen "neo-colonialistic dependence on technologically developed countries".

While firmly emphasising the necessity for pursuing scientific and technological progress, Dr Savić said that it is unreasonable in the present world situation to expect "technological development by itself [to] remove increasing potential differences" or to bring about the humanisation of society. Given the powerful forces which science has made available, the dangers which threaten the survival of the human race are great; and "we must be well aware that [they can be overcome] only by the conscious and active endeavours of all progressive forces". Put positively, this means that "it is necessary to organise the endeavours of all progressive forces in order to provide that scientific achievements serve the majority instead of the minority". "Science and technology [must] become the property of society as a whole", in fact and not just in words; and this requires the implementation of three general principles, namely:

"that the interests of the peoples are above the interests of individuals or particular castes;
that social responsibility for the application of scientific achievements should be enhanced;

that the developed must endeavour to contribute to the development of the economically and technologically underdeveloped in order to accelerate the process of development of the productive forces and to avoid the perilous consequences of existing and increasing contradictions".

The practical realisation of the goals envisaged by Dr Savić faces very specific obstacles (known generally in everyday English as 'vested interests') and can only be achieved in so far as informed action is based upon a thorough evaluation of the nature of such obstacles and of the effective significance of the forces available for overcoming them. It was in order to contribute towards an evaluation of this type that the week's working sessions were carried out.

I

Science and Technology as Formative Factors of Contemporary Civilisation: From Domination to Liberation

"Consider a typical country said to need 'to be developed'. It's probably of medium size, with a relatively large population; it has several natural resources that permit it a relative financial affluence and a genuine will to consolidate its political independence and to supply its economy with the means for autonomous growth. To these ends it is ready to set aside a not-inconsiderable portion of its foreign-exchange earnings in order to finance the importation of modern science and technology. And after a while this country realises that the conditions of a new dependence are being forged by means of technology transfer, the acquisition of prefabricated factories, even by means of technical assistance aimed at training the country's own experts. From the difficulties involved in setting up a nation-wide engineering establishment capable both of mastering scientific and technical imports and of preserving one's freedom of choice on the world market, the country realises that in order really to make use of the imported types of knowledge it would almost have been necessary to be able to produce them oneself. To use a comparison, the importation of science and technology acts rather as a drug upon which the country becomes dependent, and not as a form of nourishment for autonomous development.

"Consider the plight of a 'Third World' student in a European or American university. He's caught in the paradox of having to learn and to think in mental categories which he feels to be unsuited for dealing with his country's problems, but which he is nevertheless unable to cast aside, since his own culture (which was not able to face the trials of modernity freely and of its own accord) has not provided him with adequate alternative ones. This student develops an internal conflict, the difficult resolution of which may paralyse his possibilities for intellectual and cultural identification, rather than measuring their steady growth. At best, he comes to know what the Sudanese sociologist, poet and

I Science and Technology as Formative Factors

political figure Mohî al-Dîn Câber has termed 'cultural illusion'. His ambivalence is not productive, and there is the risk that he may slip into an attitude of cultural passivity which alone will make it bearable.

"Consider, almost everywhere in the world, what is called the 'traditional peasantry' and the regions where this peasantry predominates. The conditions in which it lives — and in the first place the economic conditions — are now actually *destroying* the meaningfulness of its life and labours. For example, [one researcher] found in Algeria (in 1967) that the revenue derived from 300 days of agricultural labour on a three-and-a-half hectare plot of land was equal to wages realised after 75 days of work in a nearby zinc mine and that this sum was equal as well to the *savings* accumulated in 40 days by Algerian workers who had emigrated to France. In these conditions the working of the land becomes a sort of absurd occupation, and it is lived as such by the peasants. It is significant, for instance, that in a census carried out at the same time as this study, 75 per cent of the Algerians questioned declared themselves to be unemployed, although only a third of the male population was really affected by unemployment. A recent report about Thai peasant families selling their children to Bangkok contractors emphasised the double absurdity of miserable urban wages passing for fortunes when compared with the monetary funds of the peasants. At the extreme of absurdity, there is often passivity, paralysis, death. — That was the impression the French province of Limoges evoked for a certain Egyptian theatre director. This was a man well acquainted with the deep misery of the Egyptian countryside, but who had not found there the fatalism and acceptance of death which he seems to have encountered in Limoges.

"Let me give two more examples from life in France, which happens to be my own country. Last summer thousands of hectares of forest in Provence burnt down. Modern technology (aeroplanes, 'canadairs', motorised columns . . .) was unable to check the disaster. If we can believe the writer Rezvani, who is a devotee to this forest, the tragedy was at least partially due to the lackadaisical attitude of the local authorities, who at the smallest alert, rather than relying on their own forces, called for the intervention of an external agency represented by heavy technology. And if one again encounters passivity and dullness here, it is because the Provençal forest is gradually being emptied of its 'traditional' population as the soil is snapped up by real estate agents and the owners of secondary residences or frozen in the land-speculations of European banks. . . . The traditional population had a know-how and techniques of fire-prevention (such as clearing off the underbrush during the winter) which was based on an intimate acquaintance with the terrain, on an almost direct fusion of the human inhabitant and the environment; such know-how allowed one to determine exactly when and where fires had to be lit for burning off the underbrush, how to keep control of such fires, etc. Heavy modern technology isn't successful in replacing this detailed knowledge formerly possessed by shepherds, hunters and woodsmen, and this is not simply a question of the means employed.

"There is a sort of subtle relation between modern techniques and social passivity that one finds in evidence with the 'programme Massif Central', inaugurated in 1975 for the purpose of checking the decline of the French province of Auvergne, which is suffering from a huge rural exodus, a depressed agricultural sector and a failing wine-industry. The programme is about half-way finished; but according to Pierre Pascalon, the results are not very satisfactory, because this project for *rural* development remains in the hands of *urbanites,* because it has been organised from Paris, in terms of progress, industry, and the optimalisation of profits ... The solution is certainly not to 'dream up some Rousseauian schema in which Auvergne would become a living example of primitive purity chained to a set of outdated values'. But it is necessary to be aware of the significance of this sort of progress that is conceived somewhere else and imported from the outside. There develops in the local population a tendency to await passively for credits, plans and ideas coming from above; and given the conditions in which the population finds itself, this attitude is logical and inevitable. What's missing is 'a local responsibility that's really lived and experienced as such by the inhabitants'.

"Fatalism of course is not always fatal. Within a favourable conjuncture of circumstances, it can happen that a local population or social group manages to integrate technical modernity into a strategy of its own creation, concretised as an activism aiming at survival or development.... But this 'interiorisation' of scientific and technical knowledge seems relatively rare, since it is contingent upon a chain of events that is itself somehow a matter of chance and thus unforeseeable.

"It's not only in the traditional sectors that science and technology are involved in provoking a type of interior exile which may at times lead to the loss of personal or social identity. This is also the case at the heart of the modern system. Take the modern factory worker who is faced with a technical rationality that has become so alien and so strange to him that he can no longer place his work, no longer make any sense of it, no longer figure out his own place within the productive machinery. Take the technician working in a large-scale research laboratory who can no longer understand how his own activity fits into the execution of the general experiment that he is helping to carry out. Take the person who's ill and who's taken into the charge of a health-care system which despite the good intentions of medical personnel and hospital staff treats him as 'an object to be cured': it's the health system that does things for him; as a patient, he 'receives' medical and para-medical attention; he's 'treated' with hospital equipment (and medications), in often-times painful conditions of ignorance about his own body, about what's going on around him, about what's going to happen to him. Take the 'average consumer' urged by modernist consumerism to consume 'rationally' and 'scientifically': he feels as much excluded from a 'science' whose soundness he has no way of testing as he does manipulated by a world of fetishising advertisements — especially when such advertisements themselves are paraded in scientific and technical jargon. These, by the way, are only a few examples out of many in which a general population

I Science and Technology as Formative Factors

becomes a necessarily voiceless and passive interlocutor, an indispensable but (because of its own ignorance) disqualified witness to battles of experts... about the safety of nuclear energy plants and the Pill, about the analysis of the economic crisis, about sexuality, about pollution and the destruction of the environment, about the definitions of mental illness and mental health, etc..."
(from Y. Barel, *Paradigmes scientifiques et autodétermination humaine*)

On Monday afternoon the symposium's first working session was organised around the theme of "Science and Technology as Formative Factors of Contemporary Civilisation – From Domination to Liberation". The five position papers presented to the session and the ensuing discussion developed the theme from different points of view, but it can easily be said that each intervention sought to focus attention on the same basic questions, namely: science and technology for whom? for whose benefit? at whose service? These questions were posed in terms of options and tendencies discernible at present – in terms, that is, of science and technology as formative factors of *contemporary* civilisation; but it was generally understood that it is objective solutions put into practice now which will fundamentally determine whether science and technology play an oppressive or a liberating role in the foreseeable future. If, on the other hand, we just quoted at length from the opening pages of the position paper by Yves Barel, this is not because one intends to give exhaustive accounts of concrete cases showing the social impact of science and technology, or because one necessarily considers all of Barel's formulations to be unproblematical. The purpose was rather to stress science and technology as *formative* factors of contemporary civilisation by evoking a number of examples which highlight the extent to which their social impact touches people (perhaps ourselves) concretely in their everyday personal experience and thus forces one to pose the social functions of science as problems.

The existence of such problems and the fact that their solution can and should be a matter of responsible human choice is at times obscured by various forms of scientism which portray modern science as a sort of disembodied saving grace, a fairy god-mother with a magic wand who can conjure up instant human happiness, while at the same time (by servilely obeying the laws of Nature in order to master them) making all human decision seem ephemeral and pointless. Such a view was attacked along two main lines.

It was first of all pointed out that science and technology are the results of historically determined social activity. Their development has not been abstract but concrete and tightly bound up with given forms of society and given social needs. The uses to which they are put and the directions in which they are developed, far from being socially gratuitous, are, on the contrary, tied to very real social and class interests. Science and technology tend primarily to serve the interests of the dominant segment of the society in which they are found; and the results – positive and negative – of their development are certainly not always distributed equally to all. Historical illustration of these points was provided by Dr Tomović as he reviewed the development of modern technology since the Industrial Revolution and considered the implications of a heritage dominated by mass production, profit optimalisation, hierarchical forms of management and the abuse of natural resources. Dr Leite Lopes extended the historical analysis in order to situate the scientific and technological dependence of the Latin American countries within the general context of their continuing industrial and political dependence; and Dr Le Thanh Khoi related specific mechanisms of scientific and technical dependence to other aspects of the broad structure of cultural domination to which Third World countries are subjected. Henri Lefebvre spoke about the difficulties involved in adequately understanding new types of relations emerging on a global scale, but he stressed the continuing pre-eminence of the world market in shaping scientific and technological as well as political objectives at this level; drawing concretely on the example of the informational sciences, he considered some vital ways in which the development of new fields of knowledge is a scene of sharp social struggle. Dr Pandeya in turn pointed out that in the Third World both the natural and the social sciences can flourish only if scientists are bound closely to the people and serve the interests of the people rather than those of the transnational corporations and their agents. Dr Barel developed these problems theoretically; working from a view of the mutual interpenetration of science and society, he considered the relations between what he termed socio-epistemological paradigms and the problem of human freedom; and this brings us to our next point.

The second attack on scientism lay in discrediting the notion that scientific truth tends necessarily to bind human action to a single, narrowly 'correct' plan and thereby to do away with the need for human responsibility and critical judgement. In analysing the structure of knowledge, Dr Barel distinguished two fundamental and necessarily complementary types of rationality, namely, the mechanistic and the dialectical. He spoke of the dangers logically and historically inherent in one-sidedly pushing along with the first alone while expecting quick solutions to complex problems to be forthcoming; and, in contrast to the present-day dominance of the mechanistic approach, he observed that human self-determination and real solutions to human problems require that the dialectical method must take on the leading role. According to Dr Lefebvre, scientific truth opens up new possibilities for social practice by disclosing the necessary parameters within which human judgement and action can be effective; it thus enhances rather than negates the possibility for human responsibility *just as* it necessitates critical political struggle for differences at all levels. Drs Pandeya and Leite Lopes nuanced the obvious point that modern science and technology can be important forces in transforming conditions in the developing countries as they emphasised that only political struggles could determine whether science and technology would play a specifically liberating role for the majority of the people in the world; and Dr Leite Lopes, in particular, noted that the goal of advancing science itself gives Third World scientists an integral interest in participating in such struggles. And finally, having considered

I Science and Technology as Formative Factors

some of the negative characteristics of modern technology, Dr Tomović spoke concretely about ways and means of breaking out of contemporary technological impasses, abandoning inefficient and wasteful lines of 'development' and generally of facing the responsibility of implementing or creating better facilities for solving individual and social problems.

Despite a tendency for consensus on general questions such as the two we have just been discussing, it should nevertheless be mentioned again that opinion at the conference was not homogeneous. Based at times on objective conflicts of interests, at times on diversity of experience and at others simply on differences in approach and expression, there were a number of disputes. One of these concerned the idea of 'appropriate technology'. Elaborating on a number of themes that had been introduced by Dr Tomović, and emphasising the implications of demographic tendencies, Dr Macura argued that the technology necessary to meet the growing needs of the population of the Third World must be appropriate, in the sense of being inexpensive, labour rather than capital-intensive, energy-saving rather than energy-wasting, and egalitarian in terms of employment opportunities and satisfaction of basic needs; the Chinese experience was cited as quite positive in this regard, although not entirely so. Dr Holland pointed out ways in which the introduction of new technologies in the industrialised countries often entails increasing the difficulties in the maintenance of employment; he stressed that large-scale technological innovation is often an aspect of heightening international competition that may lead to war, but he believed that technology could solve many human problems if developed along rational lines. Dr Pandeya, on the other hand, objected strongly to the notion of appropriate technology on the grounds that in terms of international realities 'appropriate' is often equivalent to 'obsolete for the industrialised nations' and that building a national economy upon such a principle meant accepting for one's country the status of a perpetually dependent supplier of primary products that are chained to the fluctuations of the world market; Dr Pandeya said moreover that the needs of the developing countries can only be met when such countries possess an infrastructure capable of generating the best science and technology possible at a given moment. Dr Štambuk was also of the opinion that technological self-reliance is a necessary condition for securing real national independence; he admitted that he was not yet able to understand how the development of advanced science and technology could be squared economically with limited national resources, but he felt that problems concerning the development of science and technology as well as those concerning unemployment would properly have to be seen within the more general context of changes in society as a whole. Dr Pečujlić, in turn, cited concrete cases in making the point that 'alternative' technologies which take into account both productivity and human well-being can only be born from social struggle, and not from catchy slogans. But in a second intervention Dr Macura insisted on the point that developing countries must not blindly replicate the wasteful and noxious aspects of industrialisation as evidenced in the developed world today. In the light of this discussion, it was a pity that Dr Leite Lopes had been unable to present his position paper which dealt with several of these questions and which had, in particular, suggested the strategy of collective self-reliance as an example to be followed by Third World countries in the execution and maintenance of projects that would overtax the resources of a single country. Dr Leite Lopes agreed with Dr Pandeya that 'appropriate technology' is being employed as a ruse for perpetuating the present international division of labour, and he repudiated such a strategy as unacceptable. He also pointed out, however, that if 'appropriate' is supposed to mean that technology must be economically and ecologically sensible and that it must serve the interests of the whole community as opposed to those of an elite, then it is indeed laudable — but then there is no reason why it should, in the style of the World Bank, be recommended to the countries of the South any more than to those of the North.

This brings us back to Dr Lefebvre's point about the necessity of understanding the development of science and technology within a global perspective; to use Leite Lopes' formulation, until the peoples of all nations participate as equals in its on-going development, the 'vocation of universality' specific to modern science will remain unfulfilled. Despite the variety of conditions facing different peoples around the globe, despite consequent differences in strategies and tactics, it seems clear that Drs Pečujlić and Lefebvre are correct in stressing that science and technology in general become thoroughly liberating forces only when they are the objects of social and political struggles for democracy. First and foremost among such struggles today, however, are those for the liberation of the peoples of Asia, Africa and Latin America; and it is within the perspective of this priority that any technology, whether heavy or light — and however 'appropriate' to the maintenance of other interests — must be correctly evaluated.

Let us now turn to consider in more detail the various positions taken.

In his keynote address entitled *Le nécessaire et le possible dans la formation du mondial*, Dr Henri Lefebvre tried to clarify some basic ideas about the global dimensions of the world in which we live by focusing particularly on the relationship between the new informational technologies and our joint participation in the world as a whole.

By way of introduction, Dr Lefebvre pointed out that, while many of our notions about the world as such remain ill-defined, it is nevertheless clear that this world cannot itself be depicted with scientific objectivity or exactitude and certainly not according to a fixed model, that it should be understood as a process rather than as an object or thing, although even the term 'process' implies a predetermined finality that may not be at all certain. Is it maybe worth while considering the world today as "the highest stake in a life-and-death gamble? The destruction of the planet and the emergence of a global community present themselves throughout this necessary (inevitable) gamble as two equally probable and equally improbable possibilities". In such a scenario, "there would by necessity be risk, danger and adventure, and playing would put the whole —

1 Science and Technology as Formative Factors

and thus all the possibilities — at stake". Not a bad image, perhaps? It of course raises other questions, such as "who's playing? and according to which rules? without rules? who's running the game and who set the stakes?"

A critical survey of the ideas of several European philosophers and ideologists who have developed the theme of the 'world' in one way or another (Hegel, Marx, Nietzsche, Teilhard de Chardin, Heidegger, McLuhan, Brezinski and Kostas Axelos — by whom the above-cited passage was undoubtedly inspired) shows that our vision of the world is not a blank sheet; it is covered with "weird images [and] symbolisms that are optimistic at one moment and pessimistic at the next". It is, in particular, still marred by the "methodological and conceptual vice" of Eurocentrism: "people still think of what's happening in the world as simply an extension of European Logos, of the types of production and consumption that have been born in Europe"; and this fact cannot be accounted for as a matter of bad faith alone, for, despite his path-breaking work in theorising the mechanisms of the world market, "it must be recognised now that Marx himself did not escape this sort of Eurocentrism". It is necessary today "to recognise the diversity of cultures as well as that of concepts and categories, and even the way that they are employed in discourse.... Understanding the world as a process — historical if you like, but leaving behind classical historicity as defined by *a* single memory — requires that one pass deliberately beyond Eurocentrism. There is no reason to stick to the idea that the homogeneous sides of world reality are more important than the differences. There is no reason to expect a simple quantitative extension of European Logos; one should rather anticipate qualitative transformations throughout a long and profound movement."

While a thorough and concrete account of the global dimensions of life today would probably necessitate a complex systematisation of the sectorial results of the various human sciences, several dominant themes can nevertheless be isolated and their implications then traced in terms of the ways in which information is controlled, processed and used.

Certainly one of the most striking aspects of the world today is the globalisation of the State. Although this globalisation has taken the form of a multiplicity of national and multinational States rather than of the unity of a single world-State, Hegelian political rationality is still with us: "these States form a system: and analogous if not homogeneous traits are recognisable in each particular State. The world system of States does not, however, prevent extreme fragmentation of the world as a whole; nor does it prevent the maintenance of a strict hierarchy running from the smallest and most humble State up to the superpowers." "Homogeneity — fragmentation — hierarchy" is thus a set of ideas worth retaining when the global totality is being characterised; as a generalisation, it is also applicable to other fields than just politics alone. The whole system is likewise fraught with contradictions, from peaceful conflicts at one end of the scale to the various forms of war on the other. "Hence a proposition that can be set forth as a theorem: globalisation itself takes shape according to the phenomena which block it, fix it, shatter it (obstacles, conflicts and multiple contradictions)."

Counterposed to the globalisation of the State is the globalisation of the business firm embodied in the transnational (or multinational) corporation, of which IBM is the prototype. Understanding the exact interplay between these two aspects of globalisation is one of the most important, but also most difficult, tasks in understanding the total process. It is certain, however, as evidenced in the Nora-Minc Report, that these two forms of globalisation — the one political, the other economic — both interpenetrate and confront each other on the world market. "'Exercising control over information networks, the company takes on a dimension that, properly speaking, exceeds the industrial sphere: whether it likes it or not, it participates in the global empire.... The disintegration of States at times creates a vacuum that's quickly filled by IBM's spontaneous dynamism.'"

The world market is in effect another of the striking characteristics of today's world, and it continues to operate as a single system "since the 'socialist' countries have not succeeded in setting up a second market as a rival to the first". Nevertheless, "there is no theory of the world market", since experts "know only a single element, such as the monetary system". "The world market as such", however, is to be analysed in terms of "various movements that are either superimposed and linked or else divergent in space: the movement of finished products, movements of capital and of the labour force, the flow of techniques and of knowledge, and even of signs and symbols, the flow of information and the movement of so-called cultural works, etc...." But full analysis of this quickly changing reality is complicated by the ways in which "virtualities" themselves "are taken into account, and forecasting becomes operational" as individual sectors of the market are "explored and occupied according to suitable procedures". "Hence the proposition: the necessary, i.e. the world-wide extension of merchandise, of exchange-value (of their language and their logic) opens the way for and even requires the exploitation of the possible."

In such circumstances our visions of the world have been transformed: and, in particular, ideas about what constitutes 'progress' and 'development' are being re-evaluated. "In the conventional conception of historicity, time plays a determinant role"; but as the Earth's regions and even layers available for various activities (commercial, industrial, financial, cultural and military) are more and more brought into calculations and strategies, "a quantitative and qualitative alteration is taking place: space is taking on the primordial role". A new conception of causality is emerging concomitantly, and "time itself must be conceived along other than traditional lines.... Time is being localised and each place includes a time; but world time exists nonetheless. Temporality can no longer be understood according to the cycle of births and declines (Hegel-Marx-Nietzsche) but only in terms of the conflictual relation of world-strategies."

Now, if we wish to test the implications of such general themes within the realm of informatics, it will first of all be readily observable that "technological

I Science and Technology as Formative Factors

progress [within this field] reinforces but simultaneously diversifies the world communications network" and that "it tends to build up a single network by interconnections of separate nets and by integrating different types of services". It can likewise be observed that "the ideological function" (including the production and diffusion of knowledge) of traditional institutions such as schools and universities "is being increasingly transferred to communications systems" – centres which "are administratively and institutionally controlled either by the State or by the so-called corporations". Globalisation is thus proceeding; but it is globalisation of a quite specific and dangerous sort, the "primordial danger [being] the unlimited reinforcement of the State and its various managerial, repressive and ideological capacities".

At the ideological level there is at present the menace that the definition of the political arena itself is being increasingly dictated by authorities and 'experts' – that is to say, technicians and technocrats capable of programming information; this tendency favours the personalisation of political power at the same time as it enforces the marginalisation of all independent and non-programmed political thought and action. On the one hand, this development is justified as a necessity by the presentation of a highly suspect unitary theory of the field of communications which is based on the simplistic amalgamation of an area of basic mathematics, a set of technical applications which are related to but distinct from the pure science and, lastly, a given social practice of dealing with information. On the other hand, this complex itself is vaunted as heralding the birth of a new type of society – namely, a 'post-industrial society' from which critical thought will be eliminated, since the free flow of informational data alone allows all decisions to be made 'rationally' and automatically by computer. The 'model' society envisaged by the communications ideologists is a transparent one: "No shadows or dark corners and no little hiding-places in this perfect social practice. No secrets or shames, and no discretions. Socialized information will lead to a society that will be 'fully planned, where the centre will receive from each base-cell correct messages about its own particular level' so that culture and information will 'share the same structure and the same orientation' (Nora-Minc Report) by making each individual conscious of the general and collective restraints. This is not only an ideology, but rather a mythology of scientism and a dangerous Utopia...", dangerous especially since global auditing can be installed to facilitate an efficiency that will eliminate all disturbances.

"The communications ideologists present their cluster of techniques as an objective science: as a totalising activity capable of covering, controlling and managing social reality as a whole. They don't consider themselves to be interpreting data, but to be attaining true objectivity in the social sphere. They don't want to admit that they themselves are advancing and representing a political project. But isn't subordinating social and political facts to technical factors a political act? Technicising the political and social instead of socialising and politicising the technological is in my opinion a political act that is misleadingly objective: an ideology passing itself off as a science. This affirmation by no means resolves the difficult problem of the relationships between technical and social change, but it prohibits taking as the solution that which actually poses the problem."

A particular technique "in itself sets both demands and limits. A necessity is elaborated in it and by it. But despite the pretensions of certain technocrats, this necessity is not formed or constituted as a finished system. Far from it: it in fact opens diverse, even contradictory, possibilities. As for techniques alone producing a world organism, this is a pipe-dream that doesn't stand up to analysis."

"Technologies pose without resolving the essential problem that forces a choice: a political option.... The problem is expressed as follows: systems of communication and information must be examined not in isolation but within a social and political context. Either one requires social forces to adapt themselves to the new technology, which favours a vertical and centralised structure, or one bets on the intensification of social activity without losing heart in the face of 'static' and disturbing interventions, and this favours horizontal currents. Priority is either given to mass-produced anonymous messages travelling vertically, or else to circulation among social activities.... Either one opts for total integration of the world system – or else one opts for flexibility in the network. In any case, two types of society come into relief. There is conflict, and thus simultaneously the demand for one option and a dialectical movement, for it's obvious that a society which is decentralised from the informational point of view by no means excludes centrality and vertical messages. It simply relativises them."

Something should be mentioned here about the recent history of the ideas of 'the consumer' and 'consumers' rights'. For a while the movements for the protection of the rights of the consumer engendered a few illusions about replacing 'exchange-values' with 'uses-values'. Nowadays "the concept of the consumer is becoming more and more suspect, and not without reason. What about the citizen? That's a political concept. The consumer? That's an openly depoliticised functional concept. It's serving as an ideological tool for sapping at their foundation, the theory and the practice of citizenship and of the 'rights' of man and of the citizen that are the basis of democracy.

"In order to satisfy the consumer, it suffices to make all services function 'normally'. In the name of the consumer 'normal' functioning can be required – which puts the right to strike in question."

Besides the dangers inherent in the uses which the State can make of the informational systems, there are of course also those connected with commercial and cultural programming based on the results of surveys about the 'tastes' of key consumer groups (in the middle classes, of course); these results are invoked as criteria defining consumer needs in general, and "the behaviour and psycho-sociological mechanisms of consumers thus become means of domination".

It cannot be denied, on the other hand, that citizenship in many countries has long been little more than a political abstraction and that the citizen's rights have been withering away, as it were. Is it not possible that consumers' demands

I Science and Technology as Formative Factors 15

will serve as the basis for a leap beyond just business as usual, that consumers will come to demand a qualitative change. In this case, the concepts of the citizen and the consumer could again be joined together, consolidating and enriching each other.

In any case, real mastery of information "can not come from a centralising action, from a unitary structure. Such an action is based only on redundance and repetition. Paradoxically — and from the scientific point of view alone — it is surprise effects that diminish redundance. And surprise comes from below.... In order to master communications, it must be recognised that the 'base', the micro-societies, cells or pockets (territorial or otherwise), have their own activity and dynamic, a capacity for control and self-determination. The mastery of information is a problem of political democracy.

"This brings us back to the general problematic of self-management. Communications can perhaps provide another criterion besides production and the market. How can sham self-management be distinguished from the real conditions of its effectivity? What place do (or should) the base-organisms have in the production, management and consumption of information? Self-management can only enhance itself and take on a more concrete content in dealing with problems connected with communications."

"The demand of decentralisation goes far beyond the ideas of those who propose it [simply] on the basis of technological arguments. It implies a global project. Its achievement does not just require governmental decisions. It implies a genuine political action, that is, very concrete political struggles. The base will only open the way for itself by means of effective actions. The chances are great that the summits of political and State powers will only accept de-centralisation, pluralism, micro-societies and the affirmation of differences when they are constrained and forced to do so. By what? By democracy, or, in other words, by the struggle for democracy. In effect, democracy is not something static defined by stability or equilibrium; it is rather a dynamic, a movement defined by conquest and constant reconquest...

"Political struggle for differences [thus] becomes fundamental at all levels — but not without rejecting pretensions about being different or without severe critical analysis. Not just anything or just anybody!"

Dr J. Leite Lopes opened his paper, *Science and the making of contemporary civilisation*, by sketching briefly the historical development until the present of mankind's physical and astronomical images of the world. Singling out the ancient Babylonians and Egyptians for special mention, Dr Leite Lopes praised "the superb achievements obtained by ancient societies... in Asia, Africa and Latin America", although we would suggest that, when speaking of "their *mythical* approach to the study of nature", he was rather unsocial in failing to include that of the ancient Europeans as well. Despite an understanding of the fact that "the Greeks assimilated the celestial bodies to Gods", their undeniably great importance was nevertheless overrated in statements such as, "it was... the atomistic philosophers of ancient Greece who exercised perhaps the greatest influence on the modern conception of the universe" or "Aristarchus of Samos, in the third century B.C., discovered the complete Copernican system...". The developed images of the physical world and of the universe typical of mediaeval Islam and Christianity can still be attested as represented, for example, in the works of Avicenna and Dante, although for other civilisations the task of reconstruction is not so straightforward, since so many documents "were lost or destroyed... for instance in the subjugation of the magnificent pre-Columbian civilizations...". During the sixteenth and seventeenth centuries, the modern period in Europe saw the great scientific revolution in astronomy and physics as typified by the work of Galileo and the synthesis of Newton; and during the nineteenth century, "the notion of field" was developed in physics as the work of Faraday and Maxwell on electromagnetism "culminated with another great synthesis, which unified the domains of optics, electricity and magnetism". Then, "at the end of the nineteenth century, there were the discoveries of the electron and of the proton, and a collection of remarkable questions which led, on the one hand, to the discovery of the quantum of action by Planck in 1900 and, on the other hand, to the development of the theory of relativity by Einstein in 1905".

"In his work on the special theory of relativity, Einstein... achieved a great new synthesis of apparently disconnected ideas: the prejudice of absolute simultaneity was questioned, analysed and replaced by a new conception of physical space, a new entity in which ordinary three-dimensional space and time are amalgamated to constitute a four-dimensional manifold, a consequence of which is that space may generate time, energy may generate momentum, energy is equivalent to mass, electric and magnetic fields are aspects of the same subjacent variables, the electromagnetic field." "Einstein identified the gravitational field with the tensor of the space metric, physical space as described by the laws of Riemannian geometry. The machinery of this geometry led [him] to invent his equation of the gravitational field — an equation based on the notion that matter affects the curvature of space-time and that space-time acts back onto matter and determines the nature of its motion: a revolutionary concept which destroys the old notion of space as a passive stage where events take place, without affecting them....." Without going into details here, it can be observed that such developments were closely bound up both with the elaboration of the concept of a 'superlaw' which can be interpreted as providing confirmation of the impersonal nature of scientific knowledge and with epistemological formulations which predicate a relative autonomy between experimentation and intellectual creativity.

Meanwhile, in the last fifty years physics has been dominated by the discovery of atomic phenomena and by the associated theoretical development of quantum mechanics, and much research has especially been devoted to "the ultimate constituents of matter, the so-called elementary particles". The great hope in this field at present is "to reduce the different forms of observed forces, the gravitational interactions, the weak interactions, the electromagnetic forces and the strong forces (responsible for the existence of nuclei and therefore of matter) to different manifestations of certain underlying basic entities called gauge fields. This unification is an old dream... started with the attempts of Einstein to include the electromagnetic forces in the unification of gravitation and space-time geometry"; and its realisation "will constitute a great new synthesis comparable to those... mentioned earlier in this paper". And the method introduced into theoretical physics by Einstein, "the search for symmetry groups" which leave basic physical laws invariant, is still "at the root of our present-day work".

In summing up his story, Dr Leite Lopes quoted at length from Nobel Laureate Paul Dirac to the effect that: "When one looks back over the development of physics, one sees that it can be pictured as a rather steady development with many small steps and superposed on that a number of big jumps. Of course it is these big jumps which are the most interesting feature...." In a style of expression reminiscent of the work of Thomas Kuhn, Dirac continued: "The background of steady development is largely logical, people are working out the ideas which follow from the previous set-up according to standby methods. But then, when we have a big jump, it means that something entirely new has to be introduced. These big jumps usually consist in overcoming a prejudice."

Commenting on this passage, Leite Lopes remarked that "the inventive physicist finds that he has to question this prejudice and replace it by an entirely new image of nature." It should, however, be added that inventive achievements such as these require much more than mental activity alone: and this is the lesson to be learned from the second half of this position paper, which considered the prospects for science and technology in Latin America today within the context of the evolution of economic and social conditions in that part of the world.

In considering this evolution, it can of course be remarked first of all that the pre-Columbian peoples – e.g. the Incas, Mayas and Aztecs, to name only the most well-known – had achieved considerable sophistication in such fields as mathematics, astronomy, agriculture, architecture and engineering; but in the first half of the sixteenth century the general cultures of these peoples were largely suppressed or destroyed and replaced by West European cultures. However, when modern science was born in Europe during the seventeenth century, the peoples of Spain and Portugal were for a number of reasons (of which a stifling religiosity was not the least important) generally excluded from participating in this 'big jump', and this fact played an important role in conditioning the low level of science and technology in the American colonies. Nevertheless, "in spite of difficult conditions... many talented scientists did important work in many countries of our continent, especially after the second half of the nineteenth century. What is of the greatest interest to us to see is that the state of political and economic dependence of our countries could not allow the flourishing of culture and science. The colonies of Central and South America were regarded as places rich in primary materials to be exported to the expanding capitalist countries of Europe. And these, in turn, exported to the Latin American colonies their industrial products...."

"The proclamation of political independence did not change the nature of the economic system in those countries – it was rather an opening toward their domination by Great Britain. At the same time, an ideology was taking form which stated that the process of economic development was a kind of game, of free competition, in which the most intelligent and most dynamic peoples are successful." But, in fact, "political and economic domination... prevented other societies and other peoples from competing in these games.

"And inside our countries, the national ruling classes, partners of those in the dominating foreign powers, developed an ideology according to which our countries have a vocation for the exportation of raw materials necessary to the expansion of the capitalist industrialised countries."

Subsequently, in the early twentieth century, the transformation of the Latin American economies by means of import-substitution industrialisation "had as a direct consequence the importation and imitation of products and of means of production invented abroad, [that is,] the purchase of technology developed in the advanced countries". But, contrary to the cases of such 'economically sovereign' latecomers to industrialisation as Germany, Japan and the USA, "the search for manufactured products equal or similar to those which were imported led immediately to a technological dependence", since "the scientific and technical knowledge necessary to industrialisation... were incorporated in the machines and plants imported from abroad".

This situation naturally enough had its reflection in a system of education in which scientific and technical training played a relatively small part. Originally, "the absence of industries implied no need for technological and scientific research institutes..."; but "following the industrialisation process many universities and scientific laboratories were founded or further supported and developed". However, in spite of this expansion of science, culture and the university system, the industries owned by Latin American industrialists still "depend basically on imported machinery and technology".

"... Associated to foreign enterprises from which they buy equipment and technical assistance, the national industries in Latin America almost never called for technical services from the national technological institutes. In this way, Latin American universities have generally been dissociated from studies for economic projects, [and] scientists and technologists have not been called upon to help make fundamental decisions in the formulation of economic-development programmes in these countries", whereas "in the advanced industrialised countries of the world... machines and plants that are invented depend on intensive technological research", which is, in turn, "based on investigation in the fundamental sciences that is carried out in their own institutes and universities".

Given such conditions, disillusionment would not be a surprising reaction on

I Science and Technology as Formative Factors

the part of scientists and research engineers hoping to contribute to the development of their countries, but the situation has been made even blacker by government decisions in the last twenty years aiming to "base development on the implantation of affiliates of multinational enterprises". In order to justify such decisions, "technocrats utilise the myth of technology-transfer" – but it is clear that "installation of plants by multinational enterprises... does not imply any transfer of technical and scientific knowledge, [since] the imported machines are invented, designed and built abroad, and the plans for making goods locally cannot be changed by the local national engineers'. Even if we set aside the question of whether these industrial products are really the ones which our populations need", it is clear that "the capacity for technological innovation" and "for technological invention is not transferred by multinational enterprises". "The integration of most of Latin America into the economic-cultural market of the industrial capitalist nations has thus led inevitably to an aggravation of dependence" in which "science and technology have become luxury import products – sometimes locally produced by and for a few"; and it is therefore "meaningless to urge the formulation of strategies for scientific and technological development in our nations if a corresponding political strategy is not analysed and formulated for changing the economic pattern of these countries".

Nowadays, of course, there are those who insist that developing countries "must develop only the so-called intermediate technologies, leaving the fields of advanced science and technology, the so-called big science or hard science, to the industrialised nations.... But the principle that developing nations must not have access to certain fields of knowledge is unacceptable – it would be an attempt at freezing the present division of the world into rich and poor nations, at perpetuating the international division of labour.

"Of course, appropriate technologies, in the sense that they should be financially, economically, ecologically adequate and serve the ideals of improving the living conditions of the whole community, not the interests of a privileged minority, such appropriate technologies are to be recommended not only to developing nations, but also to the rich industrialised countries", who are in fact the chief offenders in "the indiscriminate burning of fossil fuels", in "the indiscriminate exportation of sophisticated equipment... to poorer countries" simply to make a fast profit, and in the "indiscriminate automation of industries and services", as if science and technology had as their goal the liberation of people from work – in order to condemn them to unemployment and deprivation.

How are developing countries, then, to support the development of 'hard science'? It must be conceded that "a given country, with its specific resources, cannot always develop an arbitrarily chosen technology", but "even the nations of Western Europe had to... pool their physicists, technicians and financial means in order to establish a high-energy physics laboratory – the CERN...." Is it not worth imitating this idea of "pooling human and material resources among nations of a given region of the world"? "In this way is not the capacity of developing countries going to be enhanced, multiplied by a significant factor, are not fields of science and technology then open to such a group of nations, each of which would be unable to develop them in isolation?"

"Science, we have been taught — and we like to repeat it — works for mankind, for the benefit of man, for the liberation of man from work. Science and technology are, indeed, so powerful as to be able to send man into cosmic space. Are they, however, not impeded from improving the living conditions of the poor and exploited masses in Africa, in Asia, in Latin America?"

"As we follow the marvellous history of the elaboration of our scientific image of the Universe, we are tempted to say that science is a unique and universal system of knowledge, politically neutral and standing above ideologies.

"Scientific laws are, of course, valid whatever the laboratory of whatever country in which you make experiments to verify them. But science is not only a catalogue of data, names and statements. Scientific research is a dynamic process which includes the interaction of the scientific community with its surroundings, with political and social forces. The motivations for research, its planning and funding are not politically neutral, for science, in forming an interpreted picture of the world, gives us instruments for changing that world."

It is the totality of results arising from the broad spectrum of research being done which constitutes the distinct domains of science, "and it would not be correct to say that this ensemble is free from social, economic and even political significance". As Dr Rosa would also point out during the fourth session, it would clearly be inappropriate to say so of the field of nuclear energy physics simply because Einstein's fundamental work on this subject had a purely theoretical motivation.

"... Contemporary science nourishes all kinds of technologies that are responsible for change in our world. ... : from the technology of food production to the technology of the production of the most dreadful and destructive weapon systems. Scientists are thus naturally incited to think about the social, economic and political consequences of scientific research, even if their own personal work involves only abstract ideas", and "scientists belonging to countries of the Third World, in particular, are naturally led to meditate on the role which science and technology may have in the making of their societies.... In a developing country whose economy is dominated by multinational enterprises, the research work carried out in national research institutes and universities does not generally have applications for the benefit of that country, since such enterprises use their own scientific and technological knowledge"; and scientists of the developing nations can thus hardly escape the conclusion that the development of science and technology in their own countries presupposes the search for a political system whose concern will be the welfare of the whole population.

"The following questions are thus appropriate in a symposium such as the present one: which science and which culture, for which project of society in which world?

"Is the aim of science and technology to liberate man or to establish a world ruled by repression of the many poor by the few rich?

I Science and Technology as Formative Factors 21

"To my mind, there can be no other answer: science and technology must liberate man, and by that I do not mean only men and women of the advanced societies — we must work for liberation of all men and women everywhere so that science will fulfil its vocation of universality and become a patrimony of all mankind."

Since we have already given a direct translation of the opening sections of the position paper by Yves Barel, *Paradigmes scientifique et autodétermination humaine,* we need only take up here with the analytical passages that followed.

It will be recalled that the incidents and situations described at the beginning of this chapter — the financial and technological dependence of Third World countries, the cultural impasses of students sent to study in the developed world's universities, the misery of the traditional peasantry, the alienation of workers and consumers in the metropolitan centres themselves — all of these were brought up in order to stress the role of science and technology in the modern world. This role is directly obvious in some cases. In others it is much more indirect and difficult to discern immediately: one might get the impression that just a bit more 'communication and understanding', just a bit more 'conviviality' and 'human warmth', will be enough to fill the gaps and avoid the tragedies. How can modern science and technologies be implicated in what are, after all, the results of human shortcomings? In the case of the forest fires in Provence, for example, the effectiveness of modern equipment 'in itself' is beyond doubt; but the equipment is paralysed simply because the area has become a 'human desert'. If one delves beneath the surface, however, in order to explore some of the "multiple mediations structuring the vicious circle of the social and the scientific", it becomes clear that the human desert of our example is itself "the result of a habitual intermeshing of socio-economic interests and the insistence with which these interests use science and technology to consolidate their power. . . ; a cultural atmosphere has been created which forces scorn on all that is not 'scientific', on all that is connected with the traditional knowledge of the peasantry or the ordinary people". And if individual scientific or technical undertakings meet with failure, this is in a certain sense because "an *entire,* global socio-economic system tries to root itself in science and things technical. . .". Or should we perhaps say — because scientific and technical terms are invoked to justify the functioning and the real priorities of the system? In any case, while it may be correct to say that science is becoming a directly productive force, "it must not be forgotten. . . that 'science' does not act *on* 'society', but that society acts *in* science, and science *in* society. The one is not 'outside' of the other."

Now in all of the examples cited above, science and technology can be seen playing a certain role in processes that tended to "dispossess individuals, groups and nations of their powers and abilities for determining themselves. . . . Science and techniques contribute to the production and reproduction of a gap between 'those who know' and those who don't, and this gap, in turn, poses a problem of power and a problem of decision. Those who have power and the determination to make use of it have only too much of a tendency to insist on the passivity, limitations, resignation, social and cultural dependence... of those over whom they have power and 'for whom' they want something..."; and abandonment of human self-determination becomes part of a new ('soft') form of the dialectic between master and slave. One is caught, as it were, in an antinomy: science and technology are necessary for liberation, but they themselves forge new links in the chains of slavery.

"Science and technical knowledge are implicated in the multiplication of social systems that one can term essentially heteronomising (i.e. based on the distinction between a Centre that decides and a periphery that is set in motion from outside)". While heteronomy is often a good principle for building machines, it is certainly "doubtful whether this is so for living systems", and even more so (we might add) for social systems. The 'crisis of representative government' provides us with a case in point: what is being stressed is not the many imperfections of the system, but the fact that the system itself is becoming a factor of social fragmentation. "It is not the weakness of the Centre that is to be considered responsible for this fragmentation (as the traditional authoritarian ideologies pretend), but rather it's the fact that representative government, when consistently carried through, sets in motion a logic of reinforcement of the Centre and eventually always takes the place of the people in deciding what they want." Science and technical knowledge likewise are incorporated within this representational field, their aim being "to represent the symbolic activity and the cultural creation of the entire population", just as politicians, judges and administrators are "supposed to represent the political and social will of the population". "In other words, the representational system makes itself heteronomising, with the complicity of science and technical skill."

Dr Barel explained that he was by no means attempting to plead in favour of undemocratic alternatives, whether past or present, but he insisted on cutting through illusions about 'formal democracy'. For "the truth is that there are both soft and hard versions of authoritarianism" and "the final effect on the cohesion and the dynamism of a society is only partially a function of this relative 'hardness' or 'softness'". Compared with the situation today, many types of government in the past which we now consider quite 'hard' left much more room for popular initiative, simply because they lacked the sophisticated means now available for intervention and control. Of course, the choice between the carrot and the stick is "not a matter of indifference for the *concrete* everyday life of the population. But in both cases the capacity and desire for self-determination is placed in question; and it cannot be excluded that the scope of heteronomy will be extended all the more in so far as this questioning is done with subtlety. And in regard to effectiveness, it would take a good deal of naïvety to suppose that strong-arm methods of social control always necessarily 'work better' than soft ones."

In this regard, it is particularly mistaken to depict science as absolutely distinct from the society in which it is practised or as possessing a character,

I Science and Technology as Formative Factors

whether neutral or beneficent, based only upon an immutable methodology. "Scientific and technical knowledge definitely pose a problem from the point of view of the current and future destiny of human self-determination"; but in the final analysis this problem is rooted in both science and society, and not in the latter alone — it is, if you will, "socio-epistemological".

The general domain of the socio-epistemological is structured by the perennial conflict of two broad approaches, which can be designated, respectively, as the "mechanical" paradigm and the "dialectical" (or "structural") paradigm. "Schematically, it can be said that the mechanical paradigm describes the totality of the movements of matter or on matter which do not result in a modification in its structure or in the 'substance' of which it is constituted.... But there are also cases in which the movement of or on matter leads to a modification of its structure, to an internal transformation of material substances.... One then is concerned with a movement which is conventionally termed structural. In short, we can say that matter either endures or is transformed."

Several remarks are in order here.

First of all, the conflict between the two paradigms is one in which each strives for total dominance of the field of representation; but nevertheless "the total victory of one paradigm over the other can never be achieved. It is always as if human thought and know-how have to confront a paradox: in order to give meaning to the world and to human action, it's necessary to *choose* one of the two arch-types or paradigms; and at the same time, it is *impossible* to choose.... But action is in itself a decision which cannot be accommodated to the undecidability of its symbolism. And nevertheless it must accommodate itself... [for] when the choice of a paradigm has been made, people still realise that the choice leaves certain problems in suspense." The polarity between the paradigms, in turn, comes to be invoked as a "symbolic strategy", as a "human ploy" for dealing with the world and the representation of the world; and in societies that are divided into classes, each class and social group necessarily seeks to impose its own socio-epistemological model, which will give priorities for thought and action.

Secondly, it should be stressed that analysis of the strategy guiding contemporary science can lead to quite mistaken conclusions unless the interplay of paradigms is studied in its complexity, and not just abstractly. For example, "it is no longer possible to say that the mechanical paradigm ignores structures and that the structural paradigm takes them into account". For "mechanism has gone very far in taking structures into consideration, and it is not on this point that it can be differentiated from the structural paradigm. Mechanism assumes that however far one may go in the apprehension or transformation of things, there always comes a moment eventually when one is concerned with a 'substance', an entity, an element which can not be penetrated, decomposed, de-structured or restructured. Somewhere there is something which *is,* which is invariable or which, if it changes, does so without 'reason' or laws.... For the sake of convenience and brevity, this bastard structuralism which springs from a discreet but determinant mechanism can be termed 'light' structuralism or structural mechanism. 'Heavy structuralism', on the other hand, "'simply'" puts into doubt the hypothesis of the existence of primordial elements which is assumed by structural mechanism; and it admits, at least at the level of conjecture, that (since many things in the world are structured) perhaps *everything* is — everything 'thus' being susceptible to change."

It seems that certain domains of theoretical physics best embody 'heavy' structuralism. Yet, despite the prestige attached to theoretical physics, it is structural mechanism which now "dominates the symbolic stage"; and of the various sciences it would seem that molecular biology is perhaps most representative of the dominant model. If, moreover, one admits the validity of parallelisms between different levels of socio-epistemological organisation, it is quite easy to see the clear correlations between the way in which the modern industrial plant functions and the way in which molecular biology characterises life: in both cases, one witnesses "the *same* model of organisation, which — by 'structurising' mechanism — *also* really and symbolically opens up the field for a mechanisation of structures that is without precedent in history. To the effect that, in a certain manner, one is witnessing an unprecedented dilation of the mechanical paradigm." Now, whereas the dialectical-structural paradigm is founded "on a genuine self-programming and self-finalisation of a given system", the various forms of the mechanical paradigm, on the other hand, are based on a "pre-programming external to the system to be regulated". We thus come back within the socio-epistemological perspective to the problem of human self-determination.

Given the current dominance of the mechanistic paradigm, human self-determination is menaced by a severe contraction of its horizons: several tendencies contribute to this contraction, and the role of science in some of them can even be considered central. The first tendency is towards the automation of technical systems in relation not to 'mankind' in general but precisely to those who *work* with such systems, and it is important to understand that the option to make workers serve machines, rather than vice versa, is a socio-epistemological choice and not just a narrowly technical one. As pointed out by Dr Tomović, this orientation was embarked upon already at the beginning of the Industrial Revolution; it underwent a qualitative development at the beginning of this century with the introduction of Taylorism, and while Taylorism as such may well be on the decline (as certain analysts claim), this is largely because its scope was limited to certain types of manual labour. In fact, Taylorism should probably be seen as a rather primitive example of "the algorithmisation of labour" (i.e. the division of the labour process into minute, detachable units, in terms both of organisation and of products), which has for some time been increasingly felt not only by productive workers, but also by service and administrative personnel as well. This algorithmisation has, in turn, brought about a deep alteration in the role played by the State in all types of social work. Additionally, it now seems that we are undergoing an extension of

I Science and Technology as Formative Factors 25

algorithmisation to activities beyond the place of employment, and this extension is carried out as a reaction to the *social vacuum* characteristic of today's 'advanced societies'. Within the centres of power, "there thus arises the idea that it is necessary to reconstruct the social fabric, to fill the vacuum and to make the population 'participate' in such a manner that the population itself will take on a measure of regulation and social control". As noted also by Dr Kawano during the second session: "There is talk of decentralisation, of participation, of community movements, and even sometimes of self-management, and a few (rather shy, we must admit) attempts have even been made in this direction. But with this 'withdrawal' of the State the only thing at stake is the division of labour, not the division of power... the centre withdraws or makes itself discreet at the level of direct action or of institutions, while hoping that this absence will be compensated by a heightened *presence* of its own *norms*."

Unlike 'traditional' ideologies and moralities, these norms are not necessarily advanced in an openly insistent fashion. At times the explicit normative message is even consciously hidden under an aura of 'neutrality' (the technico-scientific garb is especially important here, most notably for the medical ideologies), as the important points are conveyed indirectly within the 'deep structure' of the message. The 'crisis of values' itself becomes a means of normalisation with the implementation of subtle techniques for the manipulation of permissiveness and the programmed introduction of 'novelties' which in practice only strengthen the status quo. In such circumstances a colloquial phrase perhaps comes closest to the truth of present-day social control: 'they've almost got it down to a science'. According to the logic of the system, the 'traditional intelligentsia' which was directly linked to the *aims* of such control is being replaced by certain segments of the middle classes whose horizon is limited by their fascination with *means*. All of this is almost as if one were "simulating a conflict between our two paradigms by imitating the real conflict which is simultaneously hidden and revealed by the symbolic ploy of structural mechanism in circumstances in which this form of ploy remains dominant, while more and more showing its narrow limits".

In order to challenge the dominant system, "it is necessary to start from the *durable* character" of the conflict between the two paradigms and to link this to the relative pertinence of each. The challenge must focus on the following point: "as a paradoxical strategy, structural mechanism has the peculiarity of placing the equivocacy of choice and non-choice between the two paradigms at the service of a goal which *is not* equivocal, and which is the mechanisation of structures. This mechanisation is almost the antithesis of human self-determination. The new "ploy" – if it is new – can only consist then in organising a conflict of the two paradigms which will give priority to self-determination over mechanisation."

Dr A. N. Pandeya gave an oral presentation on the topic of *Imagination, insight and understanding: reflections on the culture of science in a changing world*. Dr Pandeya took as the starting point of his reflection a aphorism from

Karl Marx to the effect that "minds are always connected by invisible threads with the body of the people"; and in developing this insight, he stressed that the transformation of minds (for example, in the sciences) and the transformation of the body of the people are necessarily very closely linked together. There is a dialectical relationship between the minds and the body of any people, and "one cannot inaugurate a transformation of minds without a transformation of the body of the people" as well. Returning to the image of invisible threads, Dr Pandeya pointed out that in Sanskrit a seminal insight into the nature of things was called a *sutra,* i.e. a thread, and that seminal minds of a civilisation are termed *sutracal* or makers of threads of reflective interrogations, insights and conclusions.

In considering science as a cultural force in the transformation from domination to liberation in the world today, "we are confronted not with science in the singular, but in the plural... [for] there are three forms of scientific practice and scientific product which confront us today".

The first of these is "the science which generates technology, which then [in turn] becomes a major force for transforming the productive forces and the productive capacities of a given society". Because of the historical circumstances in which this science was developed, however, it was for a long time "completely co-opted and controlled" by the forces governing the development of capitalism. Nowadays, however, this monopolisation of the forces of science is no longer the case, since there is "an alternative socialist order" which has demonstrated new ways in which the tremendous powers of science can be tapped and utilised.

Secondly, there are "the social, human and cultural sciences developed during the revolutionary phase" of bourgeois society, which today form "the core of its ideological apparatus, utilised to sustain and legitimise its powers, its domination and its repressive functions". From the point of view of those in the Third World, this ideological apparatus has in the last 150 years increased in terms both of sophistication and of sheer range of operation. National boundaries no longer limit its operations; it has become a "global force", and developments in Washington are quickly felt in India, Africa and Latin America.

Lastly, there is what Dr Pandeya termed "revolutionising science, science for revolutionising given structures and formations." This form of science was inaugurated by Marx during the middle of the last century on the basis of a critique of the human and social sciences of the day, and it poses the problem of developing critical forms of thought and action aimed at creating "a new social, cultural and human order". This form of science is "the crux of the problem" of the transformation of the modern world, but it is also "the most ignored in all of our philosophical, sociological and scientific discussions"; in fact, "since Karl Marx passed away from the scene, this critical revolutionary form of social human science has not advanced significantly beyond where he left it".

"It is possible for people in the north of the socialist world and in the capitalist world to continue to ignore this science... because what confronts

I Science and Technology as Formative Factors

them today as the major and central problem on the agenda in their society is not so much that of a revolutionary re-structuring or transformation, as one of continued reproduction of the present order... [Yet] more than three-fourths of humanity today will not have any just possibility of moving from domination to liberation unless this neglected approach is taken up vigorously and methodically"; and in order to accomplish this task, it is more than ever necessary to reassert the necessity of establishing tightly woven cultural bonds which unite this form of scientific analysis with the life of the people.

The growth of such a culture will require a confluence of three elements. First of all, insights developed by revolutionising science must be effectively communicated in a form which can directly reach the "basic critical audience, namely, the people. You will not have a science-founded culture for revolutionary transformation unless this missing link of communicating the scientific insights for policy, for action, for strategic and revolutionary purposes is restored and strengthened." Nevertheless, this will definitely be an uphill battle, for the existing channels of communication in the first world and in most of the third world "will not permit this kind of thing to happen"; and it is likewise "very doubtful whether it could happen again" even in the socialist world. Secondly, this communication of scientific insight will only quicken the social consciousness of the people when their notion of culture itself has been redefined and expanded to include not only what might be termed their social memory, but also their social imagination — that is, their "capacity to imagine and look forward in a prospective, courageous, bold manner to conceptualise and visualise concretely... the possibilities open to them". And thirdly, a new period of cultural flourishing will only be possible in so far as "the capacity for critical appraisal and critical reflection" becomes "a socially shared capacity", "and the essence of this new culture... will be possible only if we reach a point where the critical scientific insights of seminal minds reflecting on this theme become shareable in the way in which ordinary artifacts today are shareable".

The task of fulfilling such conditions will undoubtedly be difficult, but "in the absence of these prerequisites the hope of movement from domination to liberation will remain a pipe-dream".

In introducing his straightforwardly titled paper, *Technology and society*, Dr Rajko Tomović took up a question raised by Dr Lefebvre and observed that there is actually a double mystification of the relationship between science and technology. Dr Tomović observed that "especially in the academic world, there is a trend to identify each new basic knowledge with technology" and to assume that technological innovation follows fairly automatically from scientific advance. In fact, however, the development of new technologies is, of course, a highly autonomous field of activity governed by forces quite distinct from purely scientific considerations, and technology forecasting and assessment is consequently now emerging as a new but prominent part of all national and international planning. At a geopolitical level, "there are only a few industrial powers which can generate technology out of basic knowledge", and "the speed of technological progress has increased so much that without a strategy of technological development the economic and defensive capabilities of any country may be seriously undermined". Another form of mystification of the relationship between science and technology lies in the idea that specific forms of technological development are necessitated by the nature of the scientific knowledge upon which they have been based. "This phenomenon [of mystification] is happening regularly with all advances in technology, and not just with information. It happened with cybernetics, and it is happening now with artificial intelligence, with robotics, and so on. I think that intentional mystification is a powerful manipulative tool used to divert attention from real social problems and on to advances in technology."

Perhaps the social and political functions performed by technology today can best be understood by reviewing the general conditions under which modern technology emerged. From a historical point of view, of course, the development of modern technology is a fairly recent event, taking place as it did during the middle of the eighteenth century with the advent of the steam engine and the first automatic regulator devices. Nevertheless, it is often difficult for us to imagine how much conditions have changed since that time. "In 1750 most tools used by farmers were made either by the village blacksmith or by the farmer himself. It was only about 1850 that the equipment of the farm started coming in increasing quantities from the factories"; and yet even at that time "the structure of the working force... in England was such that the number of blacksmiths exceeded the number of iron-workers by nearly 50 per cent. On the other hand, there were nearly twice as many tailors as railway employees." Despite the technological advances ushered in by the Industrial Revolution, the conscious integration of science, technology and research was "almost non-existent until the middle of the nineteenth century"; and the extension of facilities for technical education was really only undertaken during the 1850s and 1860s. "The middle of the nineteenth century may also be taken as the beginning of a proper engineering education in Europe", and passages taken from the official university documents of that time are "highly instructive of the contempt prevailing in academic circles towards engineering". From the beginning of the Industrial Revolution until the end of the nineteenth century, the domination of private forms of property and the extension of colonial rule were perhaps the most general conditions shaping the "basic social goals" which governed the development of technology. Let us consider four such goals or principles.

(1) *Mass-production.* "The fundamental idea of developing industrial production on the principle of replaceable parts is not more than 130-140 years old." Implementation of this principle led "immediately to specialisation of production, assembly lines and services. Without such an approach many of the goods such as housing, domestic appliances, cars, television sets, etc., would remain the privileges of very restricted groups. Let us not forget that even shoes and socks were once reserved just for the aristocracy. Democratisation of potential access to goods of all kinds is definitely the historical asset of modern technology".

I Science and Technology as Formative Factors

(2) *Profit optimalisation.* Although "profit motivation played the role of the single major factor affecting the functioning, organisation and management of production processes" during the period in which modern technology was developed, there are, in fact, "very few studies based on the interaction of the profit motivation and technology"; but we can observe that two of the most negative effects of this interaction have been the absurd destruction of man's natural environment and the stultifying conditions of life and labour typical of the modern world.

(3) *Use of natural resources.* "The squandering of both renewable and non-renewable natural resources has been one of the most striking characteristics of the development of modern technology to date. It requires little imagination to understand the absurdity of the idea of building megapolises in which three or four tons of steel and several hundred litres of gasoline must be consumed each month by a single family for transportation." In this respect, a clear distinction must be made between the principle of mass production and that of the consumer society: "consumer technology, including marketing and advertising, is meant to satisfy not just the average human needs, but artificially derived demands as well".

(4) *Management.* "The socio-economic conditions upon which our technological civilisation is based have left a very deep impact in this area... [and] the fundamental principle prevailing in the practice of management technologies is that of rule by authority based upon hierarchy rather than upon full involvement of all those concerned." Consequently, at a general level "our civilisation, with minor exceptions, favours rule by an élite". This is especially true at the place of work, where, despite principles of political equality, one is "ruled by "orders and decrees" issued by élites based on hierarchy, property or education"; this fact is all the more striking because "such management in organisational principles is practised equally in societies based on private ownership and those based on state ownership of factories".

"Evidently such an undemocratic treatment of human beings cannot be maintained without a vast spectrum of repression and manipulation. One of the most subtle and dangerous forms of manipulation is deeply rooted in the prevailing educational systems" which provide "the cultural background for social inequality" and which oblige students "to become part of a competitive system in which the success of one individual necessarily depends on the failure of the others". The educational system reflects the existing relations of production and tends to reproduce them; it favours not "the most 'talented' but the most ambitious: those who have the ambition to 'improve themselves socially', by accepting the disciplinary, hierarchical nature of the school...".

On the basis of this evaluation of the historical development of modern technology, Dr Tomović proceeded to consider the technological prospects for the future; and he started from the basic proposition that the technological and productive potentials of the world are at such a level that "for the first time in its history, mankind disposes of productive facilities which, in principle, do not require *discrimination* in terms of goods in order to satisfy the basic needs of mankind (clothing, food, housing, public transportation, education, health care). Such a goal is at least technologically feasible. As a matter of fact, other factors beyond technology are still preventing faster progress in this direction."

Now, it is especially in the realm of 'software' that the greatest potential for technological progress lies. While "individual creativity and heuristic approaches were the most frequent tools of technological progress in the eighteenth century", new informational technologies now make it possible "to manage, organise and implement the transfer of new scientific knowledge into practice in all fields of human activities. By mastering this most complicated transfer process, the promotion of technology becomes an organised social activity...." Dr Tomović also said that another radical change which will continue to affect the interaction of technology and society in the future lies in the fact that, "in contrast to the extreme concentration of technological power in terms of nations and regions in the past,... know-how, advanced industry, the promotion of new technologies are no longer the privileges of a few nations in the western world. All European nations are today technologically developed. In addition, nations in Asia, Latin America, Africa and the Middle East are also in possession of powerful industries and technological resources. Such a profound change in the distribution of technological power across the world together with the fall of colonial rule have laid down the basis for a global order of equality of nations" which will be able to replace the old order of domination and monopolistic control.

Moreover, "the once homogeneous socio-political system of governing now represents but a part of the global order, and socialist ideas have become... the founding principles and guidelines of many nations, both large and small". In these circumstances, technology cannot be adequately assessed in terms of profit optimalisation alone. Its significance must be evaluated in a much broader context, in the light of its implications for the urban environment, health and human rights at the place of work.

Contemporary society continues to set new requirements on technology, and the following are a few of the most important fields in which demands for new development are being felt.

(1) *Urban technology*. Palliative measures cannot resolve the problems of the urban crisis. The basic approach to urban development must be changed; emphasis must be placed, for example, on providing better opportunities for human contacts, and new technologies must aim at finding a better balance between economic constraints and real human needs.

(2) *Health-care delivery systems*. This field probably constitutes "the great area of challenge and expectation for the future." Nevertheless, "medical instrumentation, hospital management and the existing organisational institutions are still unable to accomplish the transition from centuries of "small-scale operations" to the full coverage of citizens' demands for health services. Our current

knowledge in the fields of automation, electronics, computers and telecommunications, etc., is such that, with the concurrent efforts of science, technology and organised social forces, it actually is possible to assure a much better functioning of large-scale health-care delivery systems" than is now available; but failure to give attention to problems in this area "reflects incorrect social priorities rather than the deficiencies of current technology".

(3) *Management technology*. In the past, some of the most adverse effects on the dignity of working people were due to their subordination to élitist and hierarchical forms of decision-making. Nowadays, given the proper social and professional sense of determination, the advent of modern automation and the computer makes any kind of non-creative, degrading human work obsolete; and "the basic question of today is not mass production but the *human factor*".

In conclusion it can be said that "the fundamental goal of social and technological endeavours must become the right of each man to become free, not only in terms of civil rights, but in a much deeper sense".

"... The interaction of society and technology has reached such a level that the strategy of technological development must take into consideration not only economic values, but the full richness of the human factor as well . . . It will most probably take another hundred years before the transition will be effected from profit-motivated and economic-growth-oriented development to a technology dominated by the human factor. But the transition has already started, and we are on the road of no return. . . ."

"As pointed out above, the academic institutions of the first Industrial Revolution were not very anxious to dedicate their intellectual resources to engineering and technology. This time the university is given another chance to assume a leading role in a time of transition. We should not stand by, passively watching the emergence of new relations between society and technology."

Discussion

When the discussion began, *Dr Miloš Macura* immediately injected several tangible economic and demographic considerations, beginning with the fact that the economic gap between the rich and the poor countries now stands at 12 to 1. This gap has increased by one-third in the last thirty years, and "the present world economic system is much more hostile to the poor nations than the former colonial system was". While the process of capital accumulation at a world scale was responsible for this differential pattern of economic growth, technology was, in fact, "the instrument" which allowed it to be put into effect concretely.

As a consequence, "in 1975 there were about 300 million people unemployed or underemployed in the world". Ninety-five per cent of those people live in the poor countries, and "40 per cent of the total labour force" in those countries was touched by various forms of unemployment. Meanwhile, by the year 2000, "8 million workers will be added to the current manpower of the world; seven-eights of those will be in the less developed countries". It can thus be calculated that by the turn of the century the less developed countries "will have to provide ... about one billion working places" if they are going to solve this problem.

Modern technology, of course, is a very impressive force for the transformation of the world; but in evaluating the roles which technology will play in this transformation, we must be realistic. Any institution "behaves according to the needs of those financing it", and modern technology is no exception. Since the end of World War II, there has, in fact, been a "tremendous shift in the sources of financing". The governments of the industrialised countries have become the most important source financing technological development; the large corporations taken as a whole come second, and all other sources then follow. Consequently, the aims of scientific and technological development have been threefold: "First, it was geared towards the production of armaments and military machinery. It was [also] geared towards space programmes and other prestige projects which are indeed very useful, but don't help the developing nations." And finally, "profit was the underlying consideration" for the global direction of development. ("This U Thant used to term the 'blind forces of economics'.")

It is thus not surprising to consider the main characteristics of modern technology: It "consumes a large quantity of energy. It rejects manpower ... ; it produces goods of short duration rather than ones which can be used for a long period; and, finally, it is expensive, which is especially important in regard to the less developed countries", since they can thus afford to buy modern technology only in very limited quantities. Making a point that would be taken up by Dr Furtado in the discussion of the third session, Dr Macura said that "not only the economic gap but also the technological gap ... in production is widening".

In this situation the great challenge facing science and technology today and in the future will be that of developing an "appropriate technology" whose "basic considerations" must be "the creation of a much more productive employment ... and the mass production of the goods and services – including health and education – necessary to meet the growing needs of populations in the third world". Such technologies should be inexpensive, and they should conserve energy and natural resources. "And finally, it seems to me that new technology must be devised so as to make it possible for a society to be egalitarian ... in terms of providing employment to everyone, in terms of satisfying the basic needs of the entire population, and in terms of establishing productive relations which will diminish existing social differences."

It is difficult to say whether it will be possible to develop such technologies. So far, China is one of the few countries which has experience in this direction. "At the Vienna conference I recently met my colleague from China and asked him what their experiences were with appropriate technology in that country. He said, 'we have failed at steel production and fertiliser production but have many good experiences in other fields'. Will China and other poor countries be

I Science and Technology as Formative Factors

willing to continue the development of appropriate technologies? Will it be possible to solve the purely scientific and technical questions that will allow the development of technologies which are inexpensive, labour-saving and yet productive?"

Throughout his intervention *Dr Stuart Holland* emphasised that the world seems to be coming to the end of a long economic wave, and he stressed the importance of understanding the key factors which differentiate conditions today from those prevailing in the second half of the nineteenth century. Dr Holland pointed out that the face of British capitalism described by Karl Marx in the middle of the last century was one characterised by far-reaching process innovations, i.e. especially technical innovations in the production process and a rising organic composition of capital. Such innovations, in other words, greatly increased production in existing sectors of the economy while at the same time displacing large numbers of labourers who had previously been necessary for the maintenance of a given level of production. From the 1870s onwards, however, the labour force thus displaced was reabsorbed on the basis of a "broad wave of product innovations" which gave rise to entirely new industries (the chemical-, oil- and electricity-based industries and their derivatives in transport) in most of the countries that were pursuing capitalist paths of development. New industries based on product innovations (in the automotive, aeronautical and atomic energy sectors) continued to match process innovations throughout the first half of the twentieth century, with the result that both investments and job-places tended to increase. Yet "in the present situation it does seem clear that we are at the end of a long phase of development where the industrial innovations which took place have tended to be labour-displacing and job-creating and where, in the developed capitalist countries in particular, the unemployment created in industry was largely absorbed into the service sector". With the introduction of the cluster of new technologies built around the micro-processor, however, an analysis of conditions in the United States, France, West Germany and Britain indicates that "we are now in a situation . . . where massive displacement of labour in the service sector is likely if these technologies are fully employed". This phenomenon indicates "not only a crisis in the developed capitalist countries of weakening capital accumulation, but of process innovations . . . which are not offset by major product innovations employing large numbers of people". Looming on the horizon is the threat of massive unemployment.

In response to this situation the developed part of the world "is not adapting in anything like a rational manner". A proper adaptation would be to "relate the enormous potential increases in productivity to the distribution of resources"; but, instead, the application of technology where it can create employment is seen not as an advantage but as a disadvantage — and (according to what one really must qualify as a rather dubious analysis by Dr Holland) this mentality follows from people viewing work as "good in itself because it is a means [sic] to cash earnings which are necessary for personal security or personal consumption". Surely people view personal well-being as the "good in itself", and they would welcome technical innovation if it were implemented in such a way as to enhance rather than degrade their living and working conditions!

Dr Holland thought that "in practice . . . it is unlikely that we will create work for the thousands of millions of people in the world who need that work. But we have the capacity nonetheless to resolve many, if not most, of our problems by a rational use of technology . . . . We have the capacity. We have the power, and yet it is not applied. And here I think it is very important to reintroduce the economic elements of the analysis into the context of social psychology, of social values, of ideology in the widest possible sense — not simply the framework of ideas but of values, assumptions, of presumptions, preconditions and plain prejudices of the way in which people view the world.

"What we have to avoid is divergencies that lead to major global conflict. This includes the whole area of arms expenditure, technology and defence." And a conference of this sort must especially seek "to point out . . . the complementarities, the interdependence of any feasible kind of development on a global scale rather than simply the areas of problems, the contradictions which are likely to arise".

Dr Pandeya then took the floor and pointed out that while at the economic level there is a "12 to 1 ratio between the overdeveloped and the developing world", it is even more striking to notice that in the knowledge industry, considered by itself, "the imbalance . . . between the developed North and the underdeveloped South is of the order of 30 to 1". At the present time, and according to the most conservative estimates, the place occupied by the knowledge industry in the economies of the United States, Western Europe and Japan ("the trilateral club") has undergone a tremendous growth since the end of World War II, and by the year 2000, "the share of this knowledge industry in the US economy alone would be around 75 per cent".

What have been and are the implications of this explosion? Dr Pandeya said that in 1947, at the time of Indian independence, "we in the knowledge industry were still operating in terms of our indigenous . . . free sources, textbooks and problèmatique". But today, thirty years after independence, the knowledge industry has been completely modernised and incorporated into the international division of labour; it has been fitted out with an entire distribution network providing everything from textbooks and journals to software and models for developing future options.

"The transfer of knowledge-industry technology is so wholesale that now we have the privilege of conducting our scientific education in the basic sciences, in the engineering sciences and in the social sciences in terms of nearly 8000 textbooks which are obsolete" in their original place of production, the USA.

Responding to Dr Macura's position concerning the need for appropriate technology, Dr Pandeya said that he had to "repudiate this approach totally" and that he was "sick of being told that the solution for this three-fourths of humanity . . . is to confine ourselves to the recommendation of the World Bank, to confine ourselves to certain areas of . . . basic production" (agricultural, ex-

I Science and Technology as Formative Factors

tractive minerals, etc.) and "to a continued utilisation — on a dependency basis — of surplus and obsolete technologies which now become, by a semantic trick, appropriate for us.

". . . No country that wholesale goes after this package of recommendations will ever be liberating itself from the present hegemonic centres of global political, military and economic control. It is an invitation to perpetual slavery of an ever-increasing intensity.

"We do not accept the implication that it is good for the North to keep on exponentially increasing its productivity while all of the dirty manufacturing (and other obsolete) technology would be the exclusive preserve" of the peoples in the developing countries. Experiences of the past three decades have shown that "this is no future for us"; and unless developing countries are in control of "the basic infrastructure" for generating the highest levels of modern science and technology by their own efforts, it will not be possible "to convert the rest of our garden into a human society with any future for our millions living and to be born".

Commenting in his intervention on what he termed Dr Barel's "wonderful diagnosis" of the conditions of contemporary scientific practice, *Dr Pecujlić* elaborated on some of the ways in which the individualistic norms now prevailing in this practice actually serve to furnish "ideological legitimation" for the reproduction of the dominant economic system. In medicine, for example, this "strictly individualistic approach . . . creates the feeling [and] the appearance that it is your own fault that you are ill"; and the tacit assumption always seems to be that "the system is all right". . . . "The collective conditions of work such as labour intensity, etc., and the collective conditions of urban living . . . remain out of sight in this approach to medicine", despite the fact that it is "exactly those conditions which perhaps most affect health". The "connection between medical science and research [on the one hand] and the mode of production and conditions of work, etc. [on the other] is cut off"; and this is "not by chance", but the result of "a systematic approach" which likewise pervades attitudes towards such phenomena as social mobility and access to the university.

Under an "economic system based on the profit motive or on bureaucratic power", the role of medicine is simply "to repair the human cog in the machine" and to maintain and reproduce the status quo in social and economic relations; but because the narrowly individualistic approach to medicine in fact conceals some of the most fundamental problems, it eventually "comes into a very sharp conflict with the health of the people", which cannot be preserved in this way. Opposition to this orientation consequently arises, as we can now see, for example, from the sharp demands formulated by the working class movement in Italy: slogans such as "health is priceless" and "prevention is revolutionary" express the demand for social responsibility and control over basic living conditions. Such struggles can, in turn, influence the character of technology, as we can see from the introduction of X-ray quality control equipment in the Alfa Romeo factories. Technologically, this equipment was very progressive; but in its original form it also happened to be very dangerous to the health of the people. They demanded a change, and new equipment was thus developed which incorporated remote control mechanisms, etc., and preserved quality in performance while eliminating health hazards. What conclusion can be drawn from this? That "only out of social struggle . . . can there be born a technological innovation which both promotes production and preserves health. . . . Not without that. That is my opinion on the question of alternatives."

Addressing himself to some of the questions raised by Dr Macura and Dr Holland concerning the reciprocal impact which technology and unemployment will have on each other, *Dr Vladimir Štambuk* recalled that unemployment is the product of capitalistic ways of production and that in that sense "the question of whether society will be able to employ more or less people is very much related to the kind of society which will be dominant in the future".

Concerning the question raised by Dr Barel and Dr Lefebvre about relating the changes in technology to forms of self-management, Dr Štambuk posed the question of whether a particular new path of technological development alone is to be scrutinised or whether it is the entire industrial system which has been dominant for the last 150 years that must be called into question. In any event, on the question of alternative technology Dr Štambuk agreed with Dr Pandeya that "even 'alternative technology'. . . produced elsewhere and not in the country itself . . . will perpetuate domination and will not help liberation". And although small countries may have definite problems of economic efficiency in producing new technologies, they will in the future find themselves in "a very difficult position if they fail to take a course of technological self-reliance".

Dr Štambuk also noted that there have been recent discussions in Yugoslavia about the possibility of developing a specific technology proper to the system of socialist self-management. It seems that most people who have considered this question do not think it valid to characterise a technology as specific in this sense. On the other hand, however, there is also a keen awareness that any given society or group of societies must succeed in developing an endogenous technology "really related to their cultural, ideological, political and social needs", if they are going to avoid slipping back into a "framework of domination and exploitation".

At that point *Dr Macura* once again took the floor in order to respond to the objections voiced on the subject of appropriate technology. Dr Macura said that it is necessary to come to grips with some very fundamental questions; and, appealing directly to Dr Pandeya, he asked "What would you like to do? . . . Would you like to reconstruct the world by copying capitalist development, by creating a consumer society, by establishing a technology which would once again lead to differentiation in the world? Or would you like to rethink thoroughly what should be done and design a technology which would be appropriate for a better world and for man as a human being? That is the basic issue", and it will not be solved by "copying a consumer society with high productivity, all the gadgets and all the problems which Mr Holland has . . .

I Science and Technology as Formative Factors

mentioned". Dr Macura considered such problems most pressing for countries of the Third World, because they possess "the world's main unemployment problem"; but "the developed countries should also very seriously rethink the paths of technological development upon which they have embarked".

As for the relationship between technological development and employment, Dr Macura said that he was willing to work for only half-an-hour each day, provided that income distribution is egalitarian — "not only at the national level... [but] at the international level as well".

During the next intervention, *Dr Le Thành Khôi* considered the problems of science and technology within the broader perspective of culture and cultural domination. Working from the position that culture in general is composed of four main elements (education, science and technology, culture in the narrow sense and communications), Dr Le Thành Khôi treated each of these successively, with an eye to the situation today in countries of the Third World.

First of all, in the field of education, quantitative increases in scholarisation cannot be depreciated, but the most important consideration is "the nature of the education and of the ideology which are inculcated by means of textbooks, teaching methods and the contents of educational programmes". Within this perspective, it is all too obvious that in many developing countries education unfortunately still remains not a process for enhancing indigenous culture, but a means for propagating foreign systems of values.

Secondly, mastery of modern science and technology can be analysed in regard to three distinct aspects: the production of knowledge, the diffusion of knowledge and the application of scientific and technical knowledge. "In these three areas, countries of the Third World remain dominated... because they are still not up to producing types of knowledge that are suited to their own context." Ninety-five per cent of research carried on in the world is concentrated in the developed countries, and the governments of developing countries are often content with simply importing the results of such research. Likewise, the diffusion of knowledge is often carried out through the mediation of the developed countries, and systems of research and of prices are often such that it is very difficult for Third World countries to exchange knowledge among themselves. In the light of these facts, it is not surprising that the application of knowledge is often no more than a "mechanical transfer" which is "a negation of all endogenous creativity". In general, one can say that the types of knowledge coming from the industrialised countries are riddled by ethnocentrism.

Thirdly, in the realm of 'culture' narrowly defined, it will be recalled that colonialism tried to persuade the subject peoples that they had either no culture or at best one vastly inferior to European culture. Unfortunately, many leaders in the Third World still continue to think that western culture alone is valuable, and they fail "to seek in their own traditions that which can contribute to another culture, to a new culture". Meanwhile, however, "cultural aggression from outside takes on multiple forms...".

Fourthly, there is the field of communications, which "has become an extremely powerful instrument of domination". "Sixty-five per cent of the information in the world today is produced in the United States and then spread to the other parts of the globe, where the press agencies of the underdeveloped countries often simply reissue the same messages without questioning either their exactitude or their ideological content." [To supplement Dr Le Thành Khôi's point about the power of the modern mass media, one can consult, for example, the recent two-volume work by Noam Chomsky and E. S. Hermann, *The Political Economy of Human Rights*, Spokesman, London, 1979.]

In summary, Dr Le Thành Khôi said that it did not seem possible to him that any genuine development whatever could be accomplished without that endogenous creativity which is able to rethink all problems and to seek solutions to these problems not only on the basis of one's own experience, but also by learning from foreign experiences and by adopting foreign solutions to one's own conditions.

In the final intervention of the evening *Dr Kinhide Mushakoji* said that there did indeed seem to be problems not only with technology and the uses to which it is put, but also with science itself. Probably the chief reason for this lies in the fact that demands for new technologies exert a strong influence on directions taken by scientific research, while the economic utilisation of technology is itself determined by profit motives. It is thus necessary now "to develop a new type of science", and this new science should be influenced both by what Dr Pandeya termed revolutionising science and also "by the wisdom . . . which existed and exists in the Third World, so that the Third World should not just be the recipient, but should also consider itself as an emanator of new scientific knowledge . . .".

Of course, the tremendous development of the informational technologies has greatly enhanced the possibilities both for centralisation and for decentralisation in the world today. If these technologies are, in fact, not being developed in directions which will foster forms of self-management and independence, this is because of "the power base which is orienting and twisting the whole direction" of development. Raising the theme to which the next morning's working session would be devoted, Dr Mushakoji said that in order to challenge the dominant relations of power it would not suffice simply to deplore the fact that the Third World is the periphery in the present world system. Instead it is necessary to have a much more active strategy of collective self-reliance by Third World countries.

II

Technology Generation and Transfer: Transformation Alternatives

If the first session of the conference developed the themes of globalisation and of the reciprocal interaction of science, technology and society, and if it likewise advanced the demand that the great potentials of science and technology should be integrated into social struggles for democratic rights, then this second session may be said to have followed a pattern according to which these various threads were woven together into a single design. The dominant motif of the session thus rightfully lay in defining a realistic strategy by which the underdeveloped countries — whose peoples, of course, comprise the vast majority of the population of the globe — would be able to overcome the present cruelly unequal distribution of power over the material, and especially the technological, resources of the world.

A general theoretical framework for the deliberations of this session was provided in the paper by Dr Štambuk, who noted that definitions of 'development' and 'underdevelopment' are notoriously legion, and who took the position that adequate definitions of these phenomena must be linked to a critique of existing modes of production as such. Dr Štambuk then went on to consider various strategies for scientific-technological development; and he concluded that only a form of self-reliance rooted firmly in the capacities and interests of the working people would suffice as a steady foundation for a nation's future.

As pointed out by Dr Wallerstein during the discussion, however, it is often a good deal easier to talk about self-reliance than to achieve it, since the 'global reach' of the transnational corporations is at present working to intensify an international division of labour which keeps underdeveloped countries dependent. Therefore, according to Dr Wallerstein, national strategies of development can only be realistically conceived and carried out when they are understood as so many partial contributions to transforming the present world order. This angle was taken up in Dr Ristić's key paper on the subject of collective self-reliance among developing countries: Dr Ristić portrayed the aim of self-reliance in general as the generation of indigenous skills and technologies capable of sustaining continuous increases in production; and he observed that, far from excluding each other, national and collective self-reliance are necessary complements and should be mutually reinforcing. Every country should thus undoubtedly aim at fulfilling its own basic needs to the greatest extent possible; but individual countries standing alone inevitably find themselves in a relatively weak position not only because of the power of the developed countries and their transnationals, but also because the scale of certain operations (e.g. in the realms of science and technology) may be too great or too complex to be met with the resources of any single country. For such reasons, concerted action by the developing countries is more and more emerging as a powerful impetus for revolutionising economic and political relations at the global level; and this strategy lends itself to being adapted in several areas crucial to scientific-technological development.

Dr Ristić, in fact, considered several such areas; but one in particular — namely, that of the transfer of technology — was later considered in detail by Dr Vesna Besarović. Dr Besarović began by noting that contemporary technology-transfer is usually a means for perpetuating the structure of global inequality in general; she then gave a fascinating account of the history of legal mechanisms governing such transfers and suggested ways in which these mechanisms might be changed to the advantage of the Third World. Whereas Dr Štambuk had already mentioned that the first precondition for any successful transfer of technologies is a well-defined indigenous concept of development, Dr Besarović maintained that the realisation of such a concept requires systematic regulation of transfers by Third World State-apparatuses and is also most facilitated by the joint action of developing countries in both negotiating and regulating them. During the discussion, Dr Issa accused Dr Besarović of having placed unrealistic hopes on the benefits to be gained by a reform of legal institutions, and perhaps he was rather too severe in this; but he did succeed in drawing attention to snares inherent in the present system of transfers. And, by way of a preview, it can be noted that in the fourth session Dr El-Kholy expressed reservations in regard to projected revisions of the Paris Convention governing international property rights and that Dr Silva Michelena called specifically for the erection of a special institution charged with the responsibilities of negotiating and regulating transfers of technology for the Third World as a whole.

Taking up a theme dear to Dr Abdel-Malek, Dr Despić stressed that in building up their scientific and technological capabilities countries of the South must distinguish their own priorities from those which the developed countries might like to see them implement; and Drs Abdel-Malek and Maraj emphasised the effective exercise of political sovereignty and the right for a nation to determine its own future as the top priority to be asserted in the face of the numerous forms of subjugation by which developing countries are threatened. On the international level, however, Dr Abdel-Malek spoke of the benefits accruing especially but not exclusively to relations with 'progressive' States, while Dr Maraj thought it feasible to work with parties from all sides, provided that the freedom of decision of developing countries is strengthened. At any rate, the most sober and practical assessment on this subject was probably that of Dr Despić, who noted that those who like to act as 'proprietors' are not only to be found in 'capitalist' countries,

II Technology Generation and Transfer

and that in the fields of science and technology the basic task for countries of the South is to generate interest-based arguments which will be effective in inducing all such proprietors to share what they at present regard as their exclusive property.

Emphasis on the historical dimension of scientific-technological development was provided especially by Dr Kawano, who reviewed pertinent aspects of the experience of Japan, that major latecomer to modern science and technology, whose record has been so impressive; and Dr Kawano's paper can, by the way, be fruitfully read in conjunction with that by Dr Nakaoka in the fifth session. Dr Hassan's intervention during the discussion to the present session also broadened the historical frame of reference by evoking often overlooked lessons from the history of science and technology in the various non-European civilisations and by thus discrediting the notion that science and technology are incompatible with the different cultures peculiar to those civilisations.

In his paper entitled *Conceptions of scientific and technological development*, Dr Vladimir Štambuk first called attention to the inadequacies of current notions of development and underdevelopment; he then went on to offer a critique of existing philosophies of scientific-technological development, and finally he sketched the outlines of several elements necessary for defining a feasible strategy of self-reliance.

Terms such as 'development' and 'underdevelopment' remain notoriously vague because different types of people link them to a large variety of contradictory social goals. Nevertheless, one can make a broad distinction between two general approaches that have been "generally predominant" until now. "The first insists on growth – i.e. on quantitative indicators. Its point of departure lies in the notion that it is important to produce things: a man with plenty of goods at his disposal will have his needs satisfied. This approach is based on the view that the existing socio-economic conditions . . . should not be fundamentally changed." The second approach, on the other hand, "does not neglect quantitative aspects, but it also insists on changes of quality in human relations" and in the conditions under which production takes place.

In fact, there was little talk of development in the industrialised countries before 1975. Until then, there was "at best" talk of growth. "Changes started taking place only after the grave energy crisis shook the West", and the study of development "swiftly led some theorists in the more developed western capitalist countries to the concept of overdevelopment" and to related theories of zero growth. All such theories fail to provide an elaborated concept of development as such and confine themselves, for example, to calculating per capita GNP.

On the other hand, in the approach taken mainly by scientists from the developing countries, "the notion of development is their central category, the pivot around which scientific arguments and critiques of the existing set of international economic relations are organised". Such scholars maintain that "the developing countries . . . have the right to a more rapid socio-economic development"; and they demand "more or less radical changes in the mode of production both in the highly developed industrialised countries and in the developing countries .... Even these [authors], however, do not elaborate the notion of development to a sufficient degree, nor is it a part of a coherent theoretical concept [for them]; it is rather used as a key term."

The lack of solidity of conceptions of development becomes especially manifest in considering 'underdevelopment', to which there are apparently "seven basic approaches". According to the first, "the problem of underdevelopment does not exist at all", for by following "an inevitable historical process" developing countries will one day become the same as today's developed countries. A second group of theorists approximates the first thesis, but in a negative way: "underdeveloped countries remain underdeveloped because they have not undergone the process of exploitation to a sufficient degree. In other words, if such countries were exploited to a higher degree, they would become an integral part of developed capitalism and thus reach its present developed level." A third approach focuses on the inequality of international relations under the dominance of a 'metropolis' which 'develops automatically' at the expense of a 'periphery' (e.g. Samir Amin and A. Emmanuel). Closely related to the third approach is a fourth including *dependencia* theories (e.g. A. G. Frank). A fifth approach assumes that an underdeveloped society is one which lacks the chief characteristics of a consumer society or a Welfare State. A sixth maintains that "underdevelopment is the characteristic of certain productive relations": certain Latin American analysts thus blame feudal relations of production for the low level of development on their continent, while there is a body of Western literature which "indirectly blames the socialist mode of production" for underdevelopment. Finally, a seventh group of authors maintain that underdevelopment "is an essential characteristic of the highly developed capitalist world", in which "social differences keep on increasing ..., creating so-called 'pockets of poverty'." In this view, capitalism inevitably leads to underdevelopment.

In Dr Štambuk's opinion, the concepts of 'development' and 'underdevelopment' could be determined more adequately by being linked to "a fundamental critique of the existing mode of production" and of "the very aims of industrialisation"; in other words, they should be linked to "a critique of the existing strategic civilisational orientation". "There should be no doubt that it is possible to start looking for a different form of civilisation and thus for a new mode of production. This new mode should have its roots in the theory and practice of socialism, in the experience of the developing world and in the manifest shortcomings and contradictions of capitalism. The direction of theoretical reasoning and practical action should not therefore have their base exclusively in the critique of characteristics of contemporary capitalism. The direction which should be taken does not imply merely the overcoming of the capitalist mode of production."

"The new, socialist — and thus different — civilisation has to be based on the authentic, specific characteristics of those societies which are in search of an alternative road to development." The values and goals of such a civilisation will

II Technology Generation and Transfer

in many respects have to be "not only opposed to, but also different from the existing ones ... [and] they will have to [be structured by] and have their catalyst in the unity of potentials and interests of the broadest stratum of each society: the working class.... The experience of socialist countries and of some developing countries already provides an outline of such possibilities. The diversity, the lack of coherence and the failures do not mean that there are no positive results." It is within this framework that conceptions of scientific-technological development should be discussed.

Recalling forcefully to mind the criticism of Eurocentrism advanced by Dr Lefebvre, Dr Štambuk said: "There is a fairly widespread view, which seeks its justification in the entire Judaeo-Christian culture, that technology and its development represent a peculiar feature of Western civilisation. Arguments offered in this connection attempt to prove that this peculiarity of that culture is the reason for its 'prevalence'. Social and economic 'achievements' of the West are mainly linked to the ability of that culture to develop scientific knowledge and technological solutions. Furthermore, this ability is alleged to secure the continued superiority of Western culture, and thus its own future progress and that of the rest of the world. In our view, such arguments are compatible neither with the evidence which history offers nor with the actual creative potential of today's humanity. Such insistence on technological 'super-characteristics' of Western civilisation tend to overlook the real contradictions of the modern world; more important still, it conceals and makes obscure certain solutions and roads to development which do exist. In this respect, [it is urgently necessary to] indicate the scope and possibilities of the modern world to overcome the inequalities, contradictions and exploitation which are inherent in the concept of science and technology as developed by Western culture."

Although there can be no doubt that technology and even science are becoming "directly productive forces", there are, nevertheless, several different views on what science and technology actually are; and it is therefore most important to have explicit working definitions of science and technology. In formulating such definitions, Dr Štambuk gave a generic categorisation of science, although he did not specify any distinction between modern and traditional science: "... science is a conscious social activity which has the task of creating a systematised body of knowledge ... achieved through description and explanation of social and natural phenomena. The task of science therefore is to establish regularities (social and natural) or at least to point out the facts which may help explain certain phenomena. The new knowledge thus gained has to be verifiable and in accordance with reality; briefly, it has to help establish objective truths." Technology, on the other hand, is most acceptably defined "as a multitude of techniques and modes which are the outcome of scientific discoveries, enable people to use nature in an organised manner and help them to manage social processes".

With these general considerations in mind, it can be said that various conceptions of scientific and technological development should be judged according to two criteria: (1) how well do they "actually contribute to the swift social development of the contemporary world, and especially of its less developed part"? and (2) do they or could they offer "ways and means of using everything positive and beneficial to the development of mankind"? In current theory and practice one can provisionally distinguish four approaches to scientific and technological development. These approaches might, respectively, be termed technologically optimistic, technologically pessimistic, 'appropriate' and self-reliant.

Representatives of the first approach "ultimately regard *technology as the key which can solve all social contradictions*. This view maintains that whenever grave, apparently insurmountable social problems and contradictions arise, new technological discoveries make it possible to maintain and extend the pace and the volume of production." "The most fervent advocates of this vision are most certainly the multinational companies." While one would not want to say that the role of technology in the development of society should be denied or negatively evaluated, "it is nevertheless beyond doubt that [technological] optimism of this kind is not realistic or historically justified.... Technological solutions alone cannot solve the problem of cultural development, the dilemmas which exist when the goals of social and economic development are to be defined, the problems of colonialism and neo-colonialism, or the problem of ... the creation of political systems with democratic characteristics, etc. ... Such burning issues of today's world ... have to be solved primarily through the inclusion of broad masses of producers into the [transformational] process, the masses who should be the true masters of their own fate."

A second widely spread view of the role of science and technology in development is that which perceives *technology as a negative factor in social development*. This concept has "numerous advocates in the developed world" who "blame developed technology for many negative aspects of the capitalist world", such as pollution, stratification, overnourishment and undernourishment, etc. The negative effects of modern technology as found in both the developed and the underdeveloped countries "lead these theorists to the claim that technology is generally unacceptable to humanity as a whole" and that happiness is to be found in social life organised in small communities without the technology necessary for large-scale mass-production. Such proposals are strongly advised to the Third World countries for two reasons. The first is one of "egoism" or underhanded 'generosity'. Before the 1970s very few of these theorists had claimed that the development of technology could be socially damaging: "... the egoism of exploitation prevented them from taking such a view". Now "when the elements of crisis exist", their motives are again egoistic, and "they are not worried by the fact that over three billion people have problems which can be solved only if further development of science and technology is combined with the introduction of appropriate social relations". This, of course, brings us to the second reason motivating the proposals of such theorists, viz. "their interest in maintaining the capitalist mode of production...".

The third approach to scientific-technological development focuses on the concept of intermediate technology, also known as *'appropriate technology'*. "This concept is an applied form of the view that technological innovation can solve everything." "The essence of this concept is that the developed industrial world should continue producing new technology, which should not be sophisticated, but suited to the needs of the agrarian developing countries."

Many 'appropriate' characteristics of this technology are often enumerated. However, "the existing patterns of capitalist relations are implicitly regarded as universally acceptable, and only technology is regarded as unacceptable". Like the first two approaches, this one also has its "rational elements"; but it does not "question its own rationale", and it thus fails to clarify "the purpose of creating an alternative technology".

Finally, as a strategy for coping with the problems of development in general, *self-reliance* should only in a derivative sense be considered as an approach to problems of scientific and technological development. This is because self-reliance "explores the possibility of finding new, different social solutions, rather than alternative technological solutions.... This concept ... embraces attempts to devise certain social solutions which would enable technology and science to become the true agents and participants in *different*, socialist, roads to development."

"Revolutionary attitudes are the salient features of the basic and most numerous social group in each modern society." Revolutionary changes will thus have different features, depending on whether this basic social group consists of an industrial proletariat or a peasantry; but the "basic feature" of a self-reliant policy is "that different roads to development from those in capitalist societies are sought".

Self-reliance thus "does not reject the need for scientific and technological development ... [but] it questions social assumptions on which scientific and technological development has until now been founded". According to Dr Štambuk, a proper notion of self-reliance should not include the views of those (such as Tinbergen) who advocate specific alternative/intermediate technologies as 'self-reliant'; but it should include "the use of scientific and technological solutions which already exist", including "some of the achievements of intermediate technology". In fact, "the dilemma of developed technology versus intermediate technology is an artificial dilemma; the real dilemma is *how* to create the technology which would be suitable to those who use it, while it would solve social problems facing the developing world ...". Science and technology must "solve theoretical and practical problems facing people in the developing countries when they manage their material and social resources and their society". In this respect, as explained in the fifth session by Dr Bonfil-Batalla, it is necessary to "pay more attention to the social and human sciences and to develop them in accordance with the traditions and needs of each individual society...".

"To put all this in simple terms, if science and technology are to serve the cause of social development of the developing countries, they must not be mere copies of alien schemes. They must represent a creative effort to overcome one's own social contradictions on the basis of the interests and needs of associated producers. In this way science and technology stand a chance of becoming true elements capable of helping develop a new, different civilisation. This civilisation will be free and authentic in so far as its creation is based on authentic national and human needs, on genuine capacities rather than mere transfers, on long-term goals rather than daily objectives, on mutuality and international co-operation rather than exploitation and national egoism. . . . Such a civilisation will secure social progress but also social justice, unity but also diversity, the development of science and technology but without any domination over people, international exchange but on the basis of equality. Its social pillar will be the immediate producers; they will be its inspirers and its builders."

Speaking on a topic which has in many ways fascinated Third World intellectuals during the last fifteen years, Dr Kawano Kenji presented his position paper, *Science and technology in Japanese history.* Dr Kawano pointed out that although the natural sciences and engineering were introduced from the West more than 200 years ago, it was only after the Meiji Restoration in 1868 that great official importance was attached to science and technology. Before that time there were, nevertheless, a number of pioneering advocates of modern science such as Sakuma Shozan (1811-1864), who was "assassinated because he tried to introduce Dutch science and technology in that early stage of the modernisation or westernisation of Japan, in spite of the chauvinistic nationalism of the time". Sakuma advanced the slogan 'Eastern morals and Western arts' in order "to make clear that Eastern morals should not exclude Western technology".

When Japan began to modernise, after the Meiji Restoration, she was confronted with "eminent rivals before her as models". Establishment of "the political unity and the independence of the nation" were the main tasks confronting her; and "other issues such as the political liberty of individual citizens or freedom of ideas were regarded as far less important". Nevertheless, although "the Meiji government wanted the centralisation of administrative power and invulnerable authority, . . . it took another fifteen years after the Meiji Restoration to achieve this original aim by oppressing the opposition parties and by suppressing agitation without mercy".

In the latter half of the nineteenth century, Japan "accepted Western science and technology without reserve, while she recognised the value of her oriental tradition in the realms of philosophy, morals, literature and social sciences. . .". This option was given institutional reality in the educational centres of the day. The establishment of the 'Imperial University' in Tokyo in 1878 and Kyoto in 1898 "symbolised the government's policy for the aim of higher education in Japan — namely, for the practical purposes of rearing government personnel for technological training and the medical sciences". Although the humanities were taught at the Faculty of Letters, they were "oriented towards the classics . . . [and] inclined to be apolitical, anti-modern, idealistic and moralistic" (rather like their opposite numbers in the West during the same period).

"The political implication of the establishment of the national universities for practical purposes was to reveal and support the government's position against the private universities which had been started by intellectuals and leaders of the opposition parties during the early Meiji period." Universities such as Keio and Waseda had been established "for the study of the humanities, particularly for the learning of foreign languages and other Western-oriented disciplines such as economics and political science". The Meiji government refused to employ any of the graduates of such establishments either in government service or in state-run universities. Yet, "in order to accommodate all the graduates in applied sciences", the Government "had to encourage not only public enterprises, but also private industries"; and "once private industries were firmly established, the national universities could not keep their privilege as the sole supplier of graduates to them". After World War II especially, "the government was finally forced to recognise the rationale and role of the private universities".

Between the two wars the influence of Western ideas and techniques spread to the humanities: "in particular, ... some courageous national university professors began to criticise the status quo of the social establishment openly. Inevitably there were cases of struggle over appointments to particular chairs between the Government and the universities. Japan experienced quite a number of tragic lessons of this sort before the outbreak of World War II."

"The defeat in World War II ironically brought another industrialisation to Japan. New research fields and technology originating in the US were introduced." "... National universities have now expanded in number from 19 before the war to 93 at present"; research institutes for science and technology have multiplied greatly, and engineering faculties, in particular, have flourished. New institutes for economic and business management using quantity-analysis methods developed in the US have also sprung up. "Unfortunately", however, "the new local universities have insufficient funds and personnel to promote the study of social science..."; but "the significant change in the realm of the humanities is the alleviation of government control and interference and the disappearance of taboos in research projects...".

Despite the great changes after the War, "the leadership that the central government has shown" has remained an "unchangeable factor". Throughout the occupation of Japan by the American army "the US tried to introduce the idea of decentralisation in the Japanese administration, but met direct or indirect resistance from the bureaucracy.... The evolution of science and technology in post-war Japan has been carried on under the guidance of central government authorities such as the Ministry of Education and the Science and Technology Agency."

Since the end of World War II, the rapid growth of the Japanese economy has depended on the development of huge industries based on breakthroughs in the fields of cybernetics, electronics, atomic energy and synthetic chemistry. Yet by the end of the 1960s it had become apparent that "science and technology, which have long been thought to represent the most brilliant achievements in the world, have suddenly proved to be incompatible with human beings and their societies ... [and] could kill us all. Poisoning from agricultural chemicals and medical drugs, air pollution from the petroleum industry, water pollution from synthetised fertilisers, traffic accidents and atomic plant radiation leaks, all of these are damaging our society, although they are the by-products of modern industry, of science and technology.

"In the space of a few years, our sense of values has reversed itself. Science and technology suddenly lost their brilliant status and their impact was regarded as suspect, though people still cling to the benefits they provide." People began to feel "a strong need for decentralisation and a local autonomy, as the negative view of science and technology began to prevail among the general public, and criticism and opposition increased against the centralised policies of the government."

What have been the results? Unfortunately, they seem to have suspiciously parallelled certain phenomena described by Dr Barel: "In the spring of 1979 the local elections for governors and mayors were held in fifteen prefectures and in hundreds of cities and towns. At that time all political parties and the mass media advocated the slogan 'Here comes the age of local communities'. What did this ambiguous catch-phrase mean? . . . The result of the election showed us that all of the former governors of large prefectures such as Tokyo and Osaka who had stood for a progressive opposition party were replaced by veteran administrators in charge of local problems in the central government. For them 'the age of local communities' simply meant that a local governor with a strong connection to the central government would be able to draw out more from the central funds for his local community."

This sort of "regionalisation" is "not sufficient to satisfy the real needs of the regional community". "It will not guarantee the autonomy of the local community or its inherent creativity. After a hundred years of centralisation, Japan suffers from severe damage to the identity and independence of the local community. People should therefore claim more insistently their right to a 'regional community'. Otherwise 'the age of local communities' will end up as nothing but another deceptive slogan."

At present it is noticeable that "the government economic circles and even the mass media are continually paying lip service to the issues of the 'regional communities'". The late Prime Minister Ohira, for example, envisaged building a number of small 'garden cities' throughout the country. There are, likewise, plans carried out in the name of decentralisation which seek to transfer the administration of traffic and welfare services to local governments; and many 'local community projects' are advanced as joint ventures of the regional government and groups of local businessmen.

What effects do such changes have on the present situation of science and technology? "It is well known that the introduction of huge industrial units in the American style brought economic development to post-war Japan. But it is also well known that the elaborate conglomerate cannot evolve any further, partly because of the lack of new markets, partly because of environmental pollution and inevitable accidents, but chiefly because of the shortage of resources

II Technology Generation and Transfer

and the new energy crisis. The technology needed at this moment is not that of the huge industrial conglomerates on a national scale and the know-how to operate them, but the development of 'intermediate technology' or 'small decentralised technology' which actually meets the needs of the local community and is under control of the members of the community; and the re-evaluation of techniques for manufacture and livelihood which have been fostered and handed down in a traditional community .... A radical change in the philosophy of science has been under way for the relocation of the regions of human life along the water supply routes and the reorientation of human society as an ecological entity. As for the energy crisis, an 'automatic energy plan' will be recommended to each regional community to replace the current energy-consuming technology and way of life."

"... The extreme advocation of 'anti-technology' [however] could easily lead to a total denial of the value of science and technology.... [Yet] Japan has been so deeply committed to science and technology that it is incredible to imagine her giving up her huge research projects ..., or casting away her elaborate industrial investments.

"What we can hope for the future of Japan is not a vainglorious centralised government, but a productive administration system which honours the initiative and identity of local communities; not a huge conglomerate for science and technology, but a small-scale flexible system of technology; not only devotion to an analytical and rational science, but also encouragement of wide varieties of humanities and social sciences."

Speaking on a subject with which he has been closely involved by his work for the UN in the last several years, Dr Slobodan Ristić presented a position paper entitled *Collective self-reliance of developing countries in the fields of science and technology*. The world is in many fundamental ways still marked by a persistence of "great disparities" resulting from centuries of "economic and political domination and dependence," and the very fact that approximately 95 per cent of the world's scientific and technological capacities (or 97 per cent of all resources earmarked for research and development) are still concentrated today in the developed countries has created "a high degree of dependence of the scientific and technological potentials of the developing countries on those of the developed nations".

As noted in a recent report by the Panel of Consultants on Technical Co-operation among the Developing Countries, "traditional technical co-operation, because it was part of a wrongly conceived 'development thinking' has for the most part contributed to the transfer of inappropriate knowledge from 'developed' to Third World countries, without even a minimum effort at adaptation to the specific situations of the recipient countries. In this type of one-way transfer of knowledge, technology, in particular, was considered to be 'neutral' in social terms and 'beneficial' in economic terms. Negative effects of this transfer on employment, structure of production, patterns of consumption, income distribution, culture, balance of payments and foreign indebtedness, dependency, etc., were *not taken sufficiently into account*...." According to Dr Ristić, the introduction of foreign models of management and decision-making is, in particular, "a highly delicate matter in both social and economic terms", for management is "a social process and not a method or technique"; and the "non-critical acceptance of management concepts and practices may have serious consequences on the development of the developing countries".

With the accumulation of experiences showing that "the social, economic and cultural development of a country cannot be based on imitations", the developing countries are now increasingly "emphasising the importance of self-reliance, not only as an essential prerequisite of the successful utilisation and development of national resources", but also as the basis for a substantial transformation of the present-day world. Seen as the very "cornerstone of development", self-reliance is defined not as self-sufficiency or autarky, but as an open-ended strategy within which "the indigenous capacity for autonomous decision-making" is fundamental. The opposite of rule by authoritarian decree, it embodies "a process that takes different forms in different fields, e.g. food, finance, energy, technology, etc., involving people at all levels to decide on choices and actions to be taken, minimising dependence, maximising independence and optimising interdependence". Self-reliance in the field of science and technology "implies an in-built preference for developing indigenous technology and competence to generate and use knowledge..."; its aim should be "to identify and choose from among a set of options [in order to] acquire technology, indigenous or foreign, at the best possible terms and then to blend it with indigenous competence so as to adapt, assimilate and improve it for a continuous increase in productivity".

The concept of collective self-reliance, in particular, is gaining increasing currency at the international level. Collective self-reliance does not imply closing channels of communication with the developed countries, but it does mean filling in the major gap in today's international system of communications and co-operation — namely, that between the developing countries themselves. This strategy thus seeks to develop the indigenous capacities and resources of the Third World and to promote closer mutual co-operation in all spheres of activity. Because of the "close interdependence of collective and national self-reliance", the former cannot be realised "without corresponding efforts on the part of the developing countries in attaining self-reliance at the national level".

Every country has as one of its foremost responsibilities the formulation of scientific and technological strategies which should pinpoint both the constraints on development and the availability of human, physical and financial resources necessary for scientific and technological progress. Such strategies "should be aimed at achieving national objectives such as economic growth, the development of national capacities for innovation and education, the management of resources, a guarantee of national security and a balance of payments, improvement of the quality of life and the position of man, etc.". Unfortunately, most developing countries still do not yet possess elaborated scientific and technological strategies; and they also usually fail to differentiate between scientific and technological policies as such.

II Technology Generation and Transfer

"The development of the methodological basis of scientific and technological development strategies is [thus] a significant area of co-operation among the developing countries, since there are many common features and similarities in their conditions and the resources available to them. On the other hand, the availability of technological achievements ... and indigenous research in developing countries is highly valuable and very often suitable to other developing countries." Another area of vast possibilities for mutual exchange lies in the domain of management. In the light of the especially noxious effects of importing management concepts uncritically from the industrialised nations, "there is obviously an urgent basis *to create an indigenous concept* and methodological basis of *management* in the developing countries, along with [making] a critical use of the achievements of the developed countries".

It is nowadays estimated that "only 3 per cent of research and development activities [in the world] are devoted to the specific problems of the developing countries"; this is, of course, not unrelated to the fact that "the share of the developing countries in total expenditures for R & D (estimated at 96.5 billion dollars in 1973) was a mere 2.8 per cent, compared with 58 per cent accounting for the activities of the USA and the USSR". Despite the fact that world research and development activities are experiencing rapid growth, "the nature of world-accumulated knowledge is decreasingly relevant to the developing countries"; and, in fact, "the development of research and development activities is quite frequently in contradiction with the interests of the developing countries. While a billion dollars has been allocated to research and development of synthetics production, little or nothing at all has been done to advance tropical agriculture and to intensify the use of raw materials." And simultaneously "nearly 45.5 per cent of total expenditure has been allocated for military research and development". At present, "the developed countries ... dictate the trends of research and have subordinated science and technology to their interests...; and it is obvious that the transformation of present [world] relations is out of the question unless a more decisive influence by the developing countries is exerted on the ... trends of scientific and technological research in the world.

"A similar position in regard to the potentials, needs, natural and climatic conditions in many developing countries calls for *joint research programmes and a wider circulation of knowledge and experience among the developing countries*. Collective self-reliance is the only way to develop research programmes based on alternative strategies providing for the interdependence of existing world achievements and local conditions in the developing countries. A pooling of efforts makes possible a rational use of imported knowledge and technology, which are not lacking in the developing countries and for which immense resources are spent...."

Strengthened scientific and technological infrastructures, their closer liaison with production systems and a capacity to meet the demands arising from particular needs and objectives are general requirements for prosperous social and economic development. "National efforts are being made in this direction, and bilateral and multilateral support given; but the pace of change is slow. The process is a highly complex one and closely related to a series of objective constraints." Of course, "*scarcity of financial resources* in developing countries is the key limiting factor in expanding mutual co-operation in different fields, including science and technology". Apart from this, however, one can distinguish several problem areas endemic in attempts at establishing technological infrastructures in the developing countries. First of all, the dependence of research institutes on the State "is natural under the conditions of the undeveloped economy and social services", but "the consequences are serious with respect to the promotion of scientific and technological research.... Motivation and stimulative measures for achieving practical results are in practice negligible, and the interest of institutes in long-term co-operation with the economic sector and other users insufficient...." Secondly, owing to a lack of indigenous knowledge and experience with material resources, the majority of institutes in the developing countries resort to developing their endogenous potential and spreading knowledge through bilateral and multilateral assistance.... [This is] very often indispensable, but it creates a basis for a more lasting dependence...." Thirdly, national R & D facilities in the developing countries are rarely located in large-scale technological systems, and they are thus severely hampered in selecting and adapting imported technical skills and equipment; they tend, in turn, to remain uninvolved in areas of applied research. Lastly, "consultancy and engineering activities, as a bridge between scientific research and practice, are today very important factors both in the utilisation and development of indigenous resources and in international expansion". Developing countries, however, remain "almost exclusively *users* of these services". While there is "a slow acceptance of the consultancy approach", most personnel "prefer to be involved in academic research"; and the small number of existing indigenous organisations are "strongly influenced by those in the developed countries".

The implementation of collective self-reliance for the establishment of scientific and/or technological infrastructures is itself a very delicate task; "it can be a 'bureaucratic' act which usually results in the development of 'supra-national' institutions producing their own programmes and methods of work without taking into account the interests and needs of their founders". Therefore, according to Dr Ristić, "scientific and technological co-operation among developing countries should begin with joint programmes and projects.... Long-term interinstitutional activities of national research, consultancy and engineering organisations should be the basis and by all means the most popular form of co-operation among developing countries.... [They] may also serve as a basis for the establishment of a network of institutions of the developing countries for joint programmes in specific fields at the sub-regional, regional and interregional levels."

The accelerated development of higher education in the developing countries is an important prerequisite of their social, economic and technological transformation. Yet, while, in general, "developing countries possess a far broader educational base than the developed countries had at their disposal in the early

II Technology Generation and Transfer

stage of their industrialisation", phenomena such as cultural imperialism and the brain-drain greatly complicate matters in this field. On the other hand, a "categorical demand" has now arisen around the entire world for "a radical reform of the university system and the educational process". Co-operation among developing countries here is of special significance: "by means of joint research, exchange of experiences, exchanges of teachers and students and other forms of co-operation, it is possible to promote the concept of a new university adapted to the dynamic developmental needs of countries striving for the respect of their cultural identity and for a more equitable position in international relations."

Generally speaking, "collective self-reliance should contribute to the strengthening of the negotiating position of the developing countries in changing present inequitable relations in world science and technology". The developed countries are, of course, "equipped with their own international machinery (OECD, EEC, Comecon, EFTA)", and they "will be reluctant to abandon their monopolistic position in development and the transfer of technology and knowledge". The developing countries, on the other hand, "are gradually building up their own machinery for mutual co-operation and for strengthening their negotiating position with the developed countries"; mechanisms for these purposes include the Group of 77, the movement of non-aligned countries and various regional economic integration groups, etc. Under present conditions, however, developing countries are mainly strengthening their position "by making use of the UN development system and by organising expert meetings". The UN Conference on Technical Co-operation among Developing Countries, for example, was held in Buenos Aires in 1978; and its programme of action was adopted by both the developed and the developing countries. The implementation of such a programme is predictably "slow" and encounters "difficulties and opposition". While the total expenditure on international technical co-operation today surpasses the sum of 3 billion dollars, the share of developing countries accounts for only 4 per cent of this. Of the 800 million dollars devoted to various assistance projects within the UN development system, the mutual exchange of experts from developing countries accounts for "only 27 per cent, while subcontracting of consultancy organisations diminishes to 6.2 per cent and [that of] equipment to 2.5 per cent". Thus, a "change of policy on the part of [the international] development and financing institutions is required in order to expand the financial base of joint development undertakings by developing countries". The Third World countries, however, must take the initiative in elaborating detailed programmes and projects, for it is clear that significant achievements will only result from the organised efforts of these countries themselves.

Similarly, developing countries must work for the regulation of technology transfers by means of international legal instruments, but they must not lose sight of the fact that "these international instruments should be complementary to the national legislations of developing countries". And really enduring changes in the international system of technology transfers will ultimately be brought about only as a result of changes in the general system of world relations.

Legal aspects of the transfer of technology in modern society was the title of the position paper by Dr Vesna Besarović, from the Faculty of Law at the University of Belgrade. In situating her topic, Dr Besarović noted that the transfer of technology carried out nowadays between developed and developing countries is part and parcel of the general conditions of inequality prevailing in the world. Whether or not one agrees with her statement that "the fundamental cause of the existing differences is the unequal distribution of scientific and technological knowledge in the world, enabling a limited number of countries to make rational use of their natural wealth", it is, nevertheless, obvious that "inequality in this field plays an essential role in perpetuating such existing differences". All of the developing countries taken together contribute only 7 per cent to world industrial production; they are held in the position of being suppliers of raw materials, "having to import almost the whole technological basis of their national economy". Developing countries today on the average can only afford to spend 0.7 per cent of their GNP on research and development, while the much richer nations of the industrial world devote from 1.3 to 3 per cent to this end; and, of the 400 000 inventions registered annually, developing countries supply only 1 per cent, while the USA, the USSR, the BRD and Japan account for 73 per cent.

For the developing countries today, "the importation of technology is an imperative of economic survival" and "represents by itself no danger for a national economy". As in the case of Japan, "the dependence of national economies on the importation of technology could be a step in bridging the gap in inventive activities... on the condition that such a transfer becomes organically implanted in domestic industrial production and stimulates local creative potentials".

At present, however, the reality of technology transfer "is overwhelmingly based on pure export from developed countries.... No more than a dozen countries and approximately a hundred multinational companies possess and control all key technology. Export and import of technology neglect the real needs of developing countries, leaving them at the mercy of the big monopolies who dominate the market." The former colonial powers have transformed themselves into exporters of technology, and they are perpetuating the dependence of the former colonies in a relatively concealed manner. There are various ways in which this can be done, but "the most common method is to neglect the establishment of a satisfactory correlation between the imported technology and the traditional culture in a broader sense.... The technology imported into developing countries is often too advanced and automatised in respect to the geophysical characteristics of the country, the demands of the market and the insufficient output of production." Rather than resolving the vital problems of the nations of the Third World, "technology [thus] becomes a power which acts in discrepancy with their basic interests".

Contrary to some assertions, "the main danger for the developing countries lies not in the ability of the supplier of technology to transform it into capital, but in the use of technology in a monopolistic manner". Within this framework, comparisons between the underdeveloped countries of today and the now indus-

II Technology Generation and Transfer

trialised countries as they were fifty or a hundred years ago ring distinctly false: "the developing countries are integrated into the world system dominated by the economically developed countries" and justified by paeans to the 'international division of labour'. "The unequal technological and economic position of the contracting parties enables the suppliers of technology to have decisive control over the determination of conditions for the transfer of technology". The owners dictate not only the type of technology to be transferred, but also such particular terms as the 'tied purchase' by which they "enlarge their profit and make the position of the receiver even more subordinate, under the pretext of an efficient transfer of technology". Because "the competition on the technology market is imperfect, prices are [likewise] decided upon by the owner", who may even offer to help cover them by means of a loan. Acceptance of such loans often "makes the position of the receiver even worse", not least of all because negotiations tend to centre on the conditions of the loan rather than on the terms of the transfer of technology itself.

When turning to consider the relevance of the present legal order for the transfer of technological knowledge, one first of all must realise that, industrially undeveloped as they were, the colonial and other dependent countries were hardly involved in matters defining industrial property rights until after World War II. Until then it was mostly "industrialists and merchants from developed countries [who] enjoyed monopolistic positions on the domestic market" (in developing countries), and consequently there was little perceived need for the registration of patents and trademarks or for national laws concerning industrial property. Later, with the attainment of national independence, most developing countries acceded to pressure to grant exclusive rights to owners of technology from the developed countries; and the adoption of the first national regulations on industrial property was strongly influenced by metropolitan interests. This influence, direct and indirect, has since increased rather than abated; and the upshot is that "the national law on industrial property, instead of encouraging domestic innovative activities and having a positive effect on the flow of foreign technology (under conditions favourable to the national economy), is, in fact, disfavouring domestic industry to the benefit of foreign technology owners from the developed countries". Attempts to modify this situation, of course, meet the opposition of neo-colonialist interests.

The situation at the level of international law is similar. When the cornerstone of the present system of international protection of industrial property was laid with the Paris Convention at the end of the last century, most of today's developing countries were colonies, "and the question of their accession to the convention was solved by application of a 'colonial clause'". Third World countries who had managed to preserve their sovereignty also acceded to the convention, acting in accord with what was considered 'progressive' behaviour at the time. Yet, although the Paris Convention had "its undeniable historical significance", even at the time of its establishment "it did not suit the interests of the underdeveloped countries". Rather, "the system of international protection of industrial property was created by the industrially developed countries, and it served as a tool for the institutionalisation of the existing monopolistic and colonial position". Yet, after World War II, "a large number of developing countries acceded to the Paris Convention automatically..., in the ecstasy of the attainment of national independence and without an estimation of the impact of [such] international conventions on their national needs". Attempts to change the Paris Convention and the body of international law based upon it today meet inevitably with opposition, because "it is backed up by the most developed countries in the world".

In general, then, "it can be stated that both national and international law are, in the present conditions, components of the institutionalisation of the existing relations based on factual inequality in the international community and that they serve as the means for new forms of neo-colonialistic exploitation". Changes in these legal institutions are thus necessary which will allow them to "serve as instruments for changing existing relations to the disadvantage of those who still occupy stronger positions in the international community".

Let us, then, consider several proposals for useful changes in the legal institutions at the national level and then at the international level. At the national level, "the [first] precondition of any 'successful' transfer of technology from the viewpoint of the importing country is an indigenous concept of the economic development of the country and of the role of technology-transfer in that development. An insight into the capabilities and needs of national industrial production, on the one hand, and of scientific and technological realities..., on the other, are the decisive factors in the creation of such a concept." A second condition of particular importance is that of "the information possessed by the country importing technology". Can that country identify the owner of the technology that is required? Does it know whether other countries have similar technologies, so that it can test the possibilities of complementary purchases? Is it aware of possible conditions for the transfer of technology, etc.?

"The very existence of legal regulation in the receiving country is in itself stimulating, because it offers a sense of security to the foreign partner and makes the process of the transfer easier altogether. The legal rules on industrial property and on the transfer of technology ought to correspond in the greatest possible measure to the needs of the national industrial and economic development...; [and] they should [also] take into account generally accepted international law principles concerning industrial property...."

"The question of the compensation of the technology supplied could be solved by legal instruments in such a way as to make it dependent on the efficiency of the technology transferred in the developing country. In such a way both the foreign and the domestic partner will be interested in the effect of the application of technology." In many contracts concerning the transfer of technology to developing countries there are provisions for loans to cover the purchase. This is "really dangerous" for the partner in the developing country, "irrespective of restrictive conditions for the implementation of the transfer". It suits the developing countries best to acquire technology by complex international business law agreements, like the agreement on long-term co-operation in production or the agreement on joint-investment of the resources of a foreign partner into domestic enterprises.

Of particular significance for the transfer of technology is the introduction of institutional measures regulating transfer. "In most of the developing countries the import of technology is not subject to [state] control at all. On the contrary, domestic enterprises are given the initiative for concluding and responsibility for implementing contracts for the transfer of technology. This is one of the most serious mistakes made by the developing countries with respect to the transfer of technology.... Governmental and other public interest organs and institutions should have an important and, in some phases of the transfer, even a decisive role."

Changing international legal regulations of the transfer of technology "is a really difficult task accompanied by much resistance in the developed countries, which are not ready to exchange their monopolistic position for relations of equality by granting the developing countries preferential treatment". Under pressure the developed countries have periodically accepted 'small concessions' and 'compromises'; but their "basic tactic" is "gaining time" by the prolongation of discussions and negotiations. "On that account the developing countries should act concertedly; they should analyse their national problems and find a common denominator ... in order to achieve appropriate arrangements on the methods and conditions for the acquisition of technology." So far, they "have not demonstrated enough understanding of the need for co-operation, but have acted independently or in small regional groupings. That only suits the developed countries." Achieving agreement among the developing countries "is not a simple task" quickly accomplished. "On the contrary, it is a gradual process of conforming needs and abilities and at the same time of bridging over differences – e.g. political, geophysical, economic, social, etc." But it is probably the speediest and most effective way for changing the rules governing the international transfer of technology.

In Dr Besarović's opinion, "the only possible way to 'cure' the existing relations of inequality existing today is to grant preferential treatment to the subjects of the developing countries and/or to the developing countries themselves". This proposal, however, has met with "considerable resistance" from the developed countries and even from some developing countries "under the strong cultural and economic influence of the superpowers". One of the most interesting questions for the developing countries at present concerns the drafting of the Code of Conduct in the International Transfer of Technology. But "a code of conduct without preferential treatment for the developing countries should be an instrument boycotted by the developing countries".

Finally, "developing countries are going to be obliged for years to come to keep on importing technology ... and to strengthen their scientific-technological base". In this situation, "the lack of an international legal mechanism which could administer transfer of technology world-wide" only strengthens the monopolistic position of the developed countries and allows them "to dictate conditions which limit the development of the scientific and technological base of developing countries".

Discussion

Dr Abdel-Malek initiated discussion by emphasising the essential connection between political power and scientific-technological development. Political power is a prerequisite of any genuine development, and "the only way to promote creativity in technology and science is by means of a vigorous development of independence, of national construction, of the mobilisation of the wide potentials at work in our nations and cultures. Unless we understand that primacy is to the political, and not to the scientific, the scientific will never take off." Thus, as can be seen from the decisions and resolutions of all meetings of the countries of the non-aligned movement, from Bandung to Havana, people in the Third World "totally reject" notions of 'appropriate' technology precisely because such notions ignore and conceal the dimensions of political power and hegemony in the world today. On the other hand, "what we have in mind when we speak of creativity in technology and science is the endogenous creativity of the masses of the population around the power of decision, independent and sovereign, democratic and progressive hopefully, to tap, as it were, the depth of the historical field of their own specific cultures and civilisations... and to bring forth new things". The case of India is a good example of this: Indian intellectuals have been of an excellent quality both before and since independence; yet "the extraordinary progress of Indian science and technology" has occurred only since 1947 "because at that time the Indian Congress leading the revolution ... created the structures and impetus to mobilise these potentials, capable of creating, not alternative, but indigenous scientific and technological creativity...". Such creativity might not be appropriate: "it might be inappropriate, but this is the process of a given take-off, viz. trial and error. The point is that unless we mobilise we can't steer clear" of obstacles on the way to real development. "As long as the power structures of the developing or underdeveloped societies..., and especially [those] in the key areas of influence, are occulted..., no hope can be entertained about mobilising the potentials of the popular masses and the intellectuals towards creativity."

"Hegemony is not just a concept." It is something concrete; and it obviously has its strongest effects on the lives of people in the developing countries and on scientific-technological conditions in those countries. In this respect, Dr Abdel-Malek spoke about what he called "the dangers, the perils of equidistance ... for Third World countries who are facing hegemonism, imperialism and colonialism". Although "not at all a partisan of strategic alliances", Dr Abdel-Malek said that "maybe we need progressive forces in advanced countries and of course progressive States more, not because we like them more, but simply because objectively for the past twenty-five or thirty years, it so happened that this sector of mankind has been of more help objectively.... For us this is an irreversible experience. We cannot go back on that by equidistantiation."

II Technology Generation and Transfer

However, "instead of equidistance in cultural and scientific things, we should have ... a panel of realistic, meaningful interrelations with everybody, absolutely without any exclusion and without any ideological *a priori*. I think it makes no sense to put ideology here. It is a problem of *Realpolitik*, not of ideology.... Maybe in one phase of the balance of world power, major countries which have been reluctant here and there to take action might take some action, which might be of some use to endogenous centres of decision and ... of creativity. Maybe. We should never close our eyes to the feasibility of equable relations." Look at China.... "After fifty years of revolution [and with] one-quarter of humanity, maybe they can open up more than other people can do"; but "we should [all] live together realistically". We should not, however, "fool ourselves by imagining that we are living in a world of angels where everybody smiles at everyone else, saying 'be good', 'take off', 'let me help you'. Things are much more complex."

"If we take this realistic approach to the prerequisites of creativity, we shall see that meaningful interaction can obtain instead of a prepostulated ecumenism, which is not a fact of life ... and I think that by stressing differences we shall see how we can relate in a meaningful manner...." In this way, it will be possible to seek a realistic complementarity in objective terms. "The way to defuse antagonistic contradictions is to seek systematically, with a cool head, ways and means to complementarity — and not just ways and means to like each other and embrace each other."

Of course, there might be differences which cannot possibly be bridged. "If we locate these, we might progress a little. If we don't, ... not much will take place.... But, happily, meaningful forces are now at work... proceeding along the path of realism towards meaningful comparatism and complementarity."

Following Dr Abdel-Malek's intervention, *Dr el-Kholy* differed with Dr Štambuk on the way in which the latter had linked technology automatically with science. Dr el-Kholy noted that "historically this is not the case, nor is it absolutely the case nowadays. Technology as the sum total of human knowledge, which is used within a particular framework, a particular set of social values and relations, to satisfy a certain social demand, is not necessarily ... applied science."

Dr el-Kholy then also warned against assumptions that "technology as such has an inner dynamic which, independently of social action, results either in certain technological innovations or in certain types of products...". Assumptions of this type basically treat the development of technology as a historical accident, whereas "innovation is after all an economic reality dictated by certain sections in society which exercise a clear choice and a direct influence on the demands made upon innovators and technologists".

Dr James Maraj next took the floor and drew the conference's attention to the conditions of the island communities of the world. Although until recently little was heard about such communities in discussions concerning development and the Third World, the islands of the world now seem to be on the brink of emerging with a new importance during the 1980s. "If it is true that non-renewable resources are almost depleted and natural resources on which the major nations depend are almost fully exploited, where shall we find the wherewithal to go on? Could it be the resources of the sea?" Certainly, these questions are not unrelated to the fact that "the Pacific — the largest of the oceans — has been rediscovered by the major superpowers (and the not-so-super) in the last three to four years".

Like other developing nations, island peoples are also "concerned with meeting basic needs, feeding, clothing and housing our peoples"; they must promote health and productivity, and they must deal with population growth. Yet, throughout their colonial experience, whether British or French, education was generally limited to preparation for a "clerkship"; and "one has only to look at our education systems even today to recognise that they are weakest in science". Thus, "the number of people in small island communities around the world who have any real knowledge of science or its possible applications is extremely small". In these conditions, "it is clear that, if we are to fully exploit the resources of our 200-mile economic zones, island communities will have to rely on technology from the more developed countries — from the more developed, not from the more skilled in exploitation. We have no hang-ups about power blocks of one kind or another. We prefer to avoid polarisations and would gladly borrow techniques from here, management styles from there, and patterns of communication from everywhere. But the choice must be ours. Freedom to choose is one of the more valued assets of independence...."

"In meetings on the law of the sea — a great deal of rhetoric has flowed about what is called the heritage of Man. We are advised that certain things should not be regarded as belonging to any particular State or nation, but they should be regarded as belonging to mankind.... In my view scientific knowledge and know-how is also part of the heritage of Man. The knowledge industry — with its closed shop approach that denies us access — seems to suggest that what I know, I keep, I hold — and guard. That is not to be shared. What kind of transformation and what kind of world will this lead to? You share my resources. I cannot share yours (which include your scientific and technological knowledge)."

While the potential of science and technology to transform the world must be acknowledged, nevertheless "major problems facing mankind — such as questions of national identity, ethnocentrism, social justice and personal freedoms — do not lend themselves to the tools and techniques of technology as we [normally] understand it. For transformation to be meaningful, these matters, too, deserve our attention. It is all too easy ... to believe that the better world we seek will necessarily be a product of science and technology [alone]."

Likewise, added Dr Maraj, "there are other things to life apart from 'economic development'". While people in island communities want to assure themselves of a dignified standard of living, they nevertheless "do not wish to change one brand of magic for another or to become sacrifices on the altar of science and technology. If dependence on science and technology is going to result in a more homogeneous humanity, we would view that as a danger, for we have no burning desire to be made in the image or likeness of anyone."

In the following intervention, *Dr André Despić* agreed with Dr Abdel-Malek

II Technology Generation and Transfer

that social and political realism is necessary in discussing the role of science and technology in the transformation of the world. It is "characteristic of the present situation ... that the globe contains a spectrum of situations wider than ever before in terms of the material level of civilisation and its technical possibilities". Therefore, it is imperative to discuss different situations separately. If discussion is limited to the promises of technological development in the world in general, misunderstandings are bound to arise, "since what applies to one situation most certainly does not apply to another". When situations are distinguished in this way, it is clear that the main problem of the foreseeable future will be "to help the major portion of the globe overcome its essential problems of survival".

Meanwhile, "in the so-called developed part of the globe, ... civilisation is facing a crisis arising from several factors evolving simultaneously. The first is population expansion; the second is the rapid exhaustion of energy sources; a third is the rapid exhaustion of material resources in general; and the fourth consists of the various forms of environmental pollution." Finding solutions to these problems presents scientific research and development with a very difficult task that is "likely to absorb their capacities for some time to come. Even if it does absorb these capacities, it is still very questionable whether we shall be able to maintain the present level of technical civilisation in the future."

In regard to the problem of transferring technologies at present available, "one has to take into account the absence of motivation to give away anything without some kind of return. The problem is that the proprietors of technologies are usually not interested in ideological, philosophical, sociological and other reasons for giving away what they possess, while those who are, including us, have no power over properties, including technology. And it is interesting that proprietors do not seem to live in capitalist countries only. ... Hence, I think that one has to work very hard on finding, defining and making known all the arguments – I should say the interest-based arguments – for faster exchange of new technologies."

Dr Despić said Dr Štambuk was correct in calling for a redefinition of the concept of development, and he agreed that a definition in terms of per capita income was "very poor". He stressed the importance, however, of having "some common basis of comparison", and he wondered whether "such a common basis couldn't be found in terms of the work that has to be put into overcoming basic material needs; in other words, in terms of productivity".

As noted previously by Dr Holland, increased productivity is posing a problem especially for the developed countries. "Technological development by definition leads to increased productivity and to reduction of needs for manpower." Dr Despić was of the opinion that "no society so far has found good ways of coping with excess labour". Yet "in a rational world" it should be dealt with "by reducing working hours while maintaining a given level of income".

Dr Hossam Issa then commented on the legal aspects of the transfer of technology. Dr Issa criticised Dr Besarović for attempting what he considered to be "a purely legal solution to the problems of transfer of technology from the developed to the 'developing countries'". In his opinion "there is no possible legal solution to the problems of the transfer of technology". The reason for this lies in the nature of the multinational corporations who are today the main agents of technology-transfer from the North to the South. First of all, there is a contradiction "between the transnational character of the activities of these corporation and the national character of the law. The national law [of a developing country] cannot control the activities of the multinational corporations, because these activities are transnational, and therefore they stand beyond the realm of competence of the national law." There is a second contradiction "between the legal concept of nationality of the affiliated company in the host country and the real nationality of the capital which controls the company. The national law of the host country considers the affiliated company as national, enjoying all the prerogatives reserved to its nationals, while it is, in fact, a part of a transnational network completely controlled by the mother company based in one or another developed country." This contradiction "can affect and even distort the concept of collective self-reliance among the developing countries", because, if such an affiliated company controlled by a transnational network but constituted according to the law of a developing country starts to transfer technology to another developing country, "this can be considered as technical co-operation among developing countries". These two contradictions "make it very difficult if not impossible to imagine a purely legal solution to the problems of transfer of technology".

Another important aspect of this problem, however, is that of the legal protection of trade-marks. "In fact, in many cases of so-called transfer of technology from the North to the South, there is no transfer of technology at all, but only transfer of trade-marks covering already standardised well-known technologies. Of course, the developing countries have to pay for these trade-marks, [and this is then] erroneously labelled as transfer of technology by the multinational corporations."

In a forceful presentation foreshadowing some of the themes to be considered in the fifth session, *Dr Ahmad Yousef Hassan* emphasised that it is important to understand the relative temporariness of present-day problems of development. The "traditional civilisations in China, Islam and India", for example, knew periods during which the sciences flourished for centuries, and they will undoubtedly know such times again. It is thus not accurate to depict the African, Asian and Latin American nations as caught in a permanently desperate condition. In this sense the history of science and technology can furnish materials which are useful for bringing home the fact that circumstances at a given point in history are not immutable. To cite only one example, it is very amusing today to consider the opinions of one of the world's first historians of science, Said Al-Andalusi, who lived over eight hundred years ago. In his book *Kitab Ṭabaḳât Al-Umam* (*The Book of the Classification of Nations*), Said was concerned, just as we are today, with classifying nations according to their level of development and underdevelopment in science. "He classified as developed nations the Indians, Chinese, Iranians, etc. And he also classified the underdeveloped nations: he cited first among them the North Europeans – if you read the description of the North Europeans, you will be astonished to hear that it is impossible for North Europeans to develop." Such a description of the conditions of nations then should remind us of the relativity of our classifications now as well: "We should really look into the history of nations and the history of civilisations in general and not limit our thinking to the prevailing situation at the time being." Certainly, more research should be devoted to the history of science and technology in the non-European civilisations.

On the question of appropriate technology, Dr Hassan thought it "dangerous" to speak of "a developed science and technology for Western countries and a special kind of science and technology for Third World countries". It is astonishing for the advanced countries to say to the developing countries, "don't develop because you will run into ecological dangers". It is certainly doubtful that the Third World countries would be heeded if they offered the same advice to the West. It is also astonishing that "advanced countries are afraid for the fate of the cultures of the developing countries. They say, 'don't develop because we are afraid that your culture will be affected. You should keep your culture, you should keep your traditions.' Why should there be a contradiction between science and technology [on the one hand] and national cultures [on the other]?"

As a result of such 'solicitude', "scientists and scholars in the Third World are being confused. They are being brain-washed, because they are being given big theories about 'appropriate' technology, about 'small is beautiful', etc.; and this is really affecting development in the Third World countries...." Likewise, it seems that in international conferences on development "people have started to ignore industrialisation;... Now is it true that we can develop without industrialisation? Just because we want to be modern, because we want to be scientific, should we now discard all theories about industrialisation? Why did the socialist countries without exception adopt industrialisation? Why are the socialist countries now themselves no longer talking about this? Will they speak to Third World countries? I wonder."

Indian MP *Dr Rasheeduddin Khan* emphasised the necessity of viewing scientific-technological development in the Third World countries within the context of contemporary global developments. By way of introduction, Rasheeduddin Khan said that since 1945 the world has arguably undergone the most radical transformation of her entire history. Within this transformation, the major global phenomenon at this time is the gigantic contradiction "between global growth, incremental growth, if you please, and regional, specific, national, endogenous development". Put more simply, this contradiction is manifested as that "between the post-industrial societies and the post-colonial societies"; and it is manifested here on at least seven counts. The first manifestation lies in the opposition between the "exponential development in science and technology, on the one hand, and the inching forward of socio-economic transformations, on the other... that is, the jet-age speed of techno-scientific transformation is matched by the ox-cart speed of transformation in Africa, in Asia and in certain parts of Latin America". Secondly, "the phenomenal expansion of world trade... is matched by stagnant islands of pre-industrial commercial patterns". Thirdly, there is "the contradiction between revolutionary transformations in communication, media and transportation, on the one hand, and the utilisation [of such facilities] by a very small portion of mankind, almost exclusively in a few of the world's advanced communities". The fourth manifestation is found in the "disproportionate consumption of energy, natural resources and services" by the few industrially advanced countries and "the increasing depletion and even exhaustion of natural wealth in the former colonies". The fifth is the "tremendous wasteful expenditure on armaments and the arms race", as opposed to the "lack of resources... available for human developmental activities". Sixth is "this increasing crescendo of emphasis on individual human rights to the utter neglect of the rights of humanity". People are faced with the suppression of their right to "life, national independence and dignity", while emphasis is placed on "artificial rights of the individual human being in a society whose structures themselves secrete inequality and violence" (e.g. the caste system in India). Lastly, there is a contradiction between urban-dominated social, political and media systems (on the one hand) and a large rural hinterland (on the other). "The conclusion is very clear. The gap between what are called the developed and the developing worlds has increased in the last thirty years much more than it ever did before. In other words, the colonial system was a more simple system than the post-colonial system. I would like to qualify this: the post-colonial system appears post-colonial in name only, for in regard to its actual stranglehold on people's minds, liberties and creativity, it is much more firm and much more sophisticated...."

The second major phenomenon in the world today is that of the "globalisation of human affairs. At a time when national sovereignties have emerged, when national identities have been created and people are struggling, fighting, laying down their lives for the maintenance of these identities, the world has become highly globalised." Globalisation of the economy, globalisation of techno-scientific culture, globalisation of value-orientations and culture have all taken place; and, in general, this "globalisation encroaches on and abridges the autonomy of the action of individual citizens in their societies". The transnational corporations are probably "the most vivid examples of the globalisation of the economy which has today reached such a point that even the advanced, post-industrial countries are in peril. Today the great United States dare not act in contradiction to the global policy perspectives of the transnational corporations.... I mean, it is very funny. How is it possible for the dollar, which is one of the world's hardest currencies, always to go up and down? It is only possible because the multinational corporations can at any given moment transfer their assets, e.g. from Los Angeles, Frankfurt or Zurich to Panama City...." Likewise, in India "the World Bank refused to advance loans unless population control went forward. Therefore, the excesses in the family planning programmes carried out under the auspices of Mrs Gandhi are ultimately traceable to the screws which the international institutions had put on India...."

II Technology Generation and Transfer 65

In the light of such pressure, as well as in the light of the powerful yet subtle forces which modern science and technology make available, it is indeed strange to the honest observer that peoples in the developing countries are advised to content themselves with natural bliss and intermediate technology. In this regard a Gandhian model of technology is at times proposed. In Dr Rasheeduddin Khan's opinion, "the Gandhian model is more in the nature of a moral imperative which one might keep in mind by emphasising that small is beautiful, that large growth alone will not do; but I doubt whether the actual [working up] of a Gandhian model into a third technology is possible...". In most of the developing countries, the problem "is really one of how to distribute justice and welfare without allowing the small, rural-urban élite to dominate all positions of power".

In the last intervention of the morning, *Dr Immanuel Wallerstein* analysed several of the disagreements which had marked the proceedings of the conference, and he suggested that several apparently contradictory positions were in fact complementary. It was first of all recalled that Dr Pandeya and Dr Macura had differed during the first session over the question of 'appropriate technology'; Dr Macura championed it and Dr Pandeya objected on the grounds that it would not lead to liberation for the Third World; Dr Macura, in turn, warned against repeating the mistakes of the developed world. Secondly, it was noted that there had been dispute about whether the locus of the problematic about scientific-technological development lay in the nature of science itself or in the specific social uses to which it was put: Dr Barel, for example, had argued that science itself is socio-epistemological and thus problematical, and other participants had, in turn, taxed this approach as involving a subtlety which might appeal to people in developed countries but which had little meaning in the technologically backward conditions of the developing countries.

According to Dr Wallerstein, the opposing positions in each of these disputes were to a certain extent the expression of a theoretical problem which had not been clearly posed and whose essential elements were thus coming to the fore as a clash of opinions. "On the one hand, we all want equality which has a flavour of sameness, and then, on the other hand, we are using as a code-theme the very legitimate and important concept of endogenous intellectual creativity, which implies the opposite of sameness. It implies that there are not only real differences, but real different potentials within this world...." Now, "presumably", the aim of endogenous creativity is to "somehow help us... on the way to equality"; and it must therefore comprise two aspects. The first is that of power, for "endogenous intellectual creativity can by its appearance and cultivation change real power relations in the world... [and] then have some impact on this issue of the growing polarisation" of the various parts of the world. The second aspect is that of interaction, of the possibility "that out of this wealth which is... the multiplicity of world civilisation, some different ideas may, in fact, emerge about very central problems like technology and science.... And [perhaps] new ways of thinking about things will, in fact, be translated into social realities."

In this connection, the concept of self-reliance raises several problems. Self-reliance is "a very nice slogan" which "has its dangers, because it implies that separate States can, in fact, transform their own situation. The reality of the world for a long time now has been that States are not real economic units." Globalisation has been under way for centuries and has entailed "a polarisation in material distribution and power distribution within the world economy... [which] is continuing. Everything that is occurring in the world today is making it bigger, not smaller ... and the issue is what kind of policies will, in fact, reverse this." Indeed, many people have already tried to reverse this polarisation, "some people for one hundred to one hundred and fifty years"; and "in fact, the net result of their very serious efforts to reverse this growing polarisation... have, in fact, increased... it, not decreased it".

According to Dr Wallerstein, no national government, no matter how powerful, can today completely control the economic conditions prevailing within its borders. The lesson to be drawn from this is that "if we want to change the economic realities of *the country* in which we live, then the only way to do that is to change the economic realities of *the world* in which we live". It is only by working towards this goal that activities for political transformation of any given country become meaningful. "... by changing policies within given countries that may have an impact on the whole world structure..., through that intermediate level we rebound back eventually on a more egalitarian world. But if we leave out the intermediate step from our own thinking, then we come into the contradiction that we have had in the last fifty years, [viz.] that our national policies aimed at reducing inequality have, in fact, increased world inequality."

III

Biology, Medicine and the Future of Mankind

On Tuesday afternoon the conference heard three presentations which, from quite distinct points of view, evoked some of the complex aspects of interaction between a society as a whole and the various individuals who together make it up. In a session entitled 'Biology, medicine and the future of mankind' one was reminded that both biological theory and medical practice have to a significant extent been marked by a peculiar fixation on the individual organism. Such fixations, however, may be said to have been outdated by developments in both fields; and – almost the same way as in physics – considerations of a more general (and in this case usually social) character must be invoked in order to properly understand and deal with phenomena which surpass the bounds of that which was formerly taken as 'typical' or 'normal'. At the same time, however, it is also clear that this fixation on 'the individual' (often connected with the name of Virchow) has by no means prevented constant depersonalisation of the doctor-patient relationship; neither has it prevented the utilisation of biological knowledge for purposes of increasing the uniformisation not only of animal and plant cultures, but also of human behaviour. After reading through the papers in this session, one might perhaps say that what they are calling for, then, is a heightened respect for individuality that is rooted both in enhanced forms of socialisation at several distinct levels and in an awareness of what are often social determinants of biological phenomena.

In the first paper Dr Ribes gave a brief sketch of the state of play in biological research in general, and he considered a few of the factors pushing this research forward. He then linked present research prospects to underlying theoretical and philosophical themes, stressing especially the 'relational' character of life itself and the essential place reserved for this 'relational' character in the course of what he distinguished as 'vertical' evolution. Dr Ribes, in turn, demonstrated that questions of social ethics are never far from the surface in the life sciences, and he suggested a series of five basic guidelines pertinent to the utilisation and application of biological knowledge. Questions of evolutionary theory were at the heart of Dr Ribes' paper; and related ones were likewise crucial to Dr Mori's presentation of the views of Japanese ethologist and anthropologist Imanishi Kinji. An important part of Prof. Imanishi's work has been devoted to an understanding of the transition(s?) from primate 'society' to human society; and his own 'bio-sociological' approach has been developed in contrast to that of Darwin and (as I learned later from Dr Nakaoka) to some of the positions taken by Engels. Dr Mori began with a quite interesting exposition linking recent developments in biological and behavioural theories with changes in political and social relations; and then went on to attempt an extension of Imanishi's 'bio-sociological' approach to evolution in order to explain the generation of conflicts within contemporary human societies. That his efforts in this last attempt were perhaps less than satisfactory was pointed out sharply by Dr Pandeya in the first intervention in the discussion. Dr Pandeya warned of the dangers of obscurantism and reductionism, which he considered inherent in this effort; he insisted on the importance of formulating concrete analyses of specifically social and political problems.

In the third position paper in this session Dr Milanović spoke about the necessity of redefining the nature and the scope of the responsibility of the physician. While stressing the continuing historical importance of the responsibility incumbent on the medical practitioner as a member of a profession, Dr Milanović stated that modern conditions now require the physician to conceive his duties in terms of a social rather than a narrowly professional obligation. It was again Dr Pandeya who added a bit of pepper to discussion on this point when he observed that the improvement of health-care systems around the world cannot be accomplished either by a redefinition of the responsibility of the individual physician or even of that of a national medical community taken by itself, because the main obstacles to effective health care are the transnational corporations dealing in pharmaceuticals and medical equipment and because these corporations are motivated by considerations of profit rather than of health. Dr Tsurumi later on agreed with this point, but, in turn, observed that in Japanese experience it seemed that physicians typically had to make the choice between serving the transnationals and serving the people. Drs Milanović and Tsurumi can thus be said generally to have put emphasis on social conditions of medical practice, while Dr Pandeya stressed one of its major economic determinants. In the last intervention of the day Dr Rakić argued forcefully that an adequate understanding and treatment of many of the most important diseases nowadays itself requires implementation of a multidisciplinary approach incorporating knowledge not only from the medical and pharmacological, but also from the human and social sciences.

In the midst of the discussion, two interventions were made which bore on the topic of the conference as a whole rather than on the particular questions raised in this third session. In an intervention scheduled originally for the first session, Dr Furtado made the theoretical point that various technologies are important not 'in themselves', but as specific means more or less effective in endowing those who possess them with particular forms of power for transforming the world. Dr Furtado also previewed positions developed in detail in the paper by Dr Vidaković and noted that since 1945 scientific-technological progress has been dominated by the arms race. Lastly, he differed with

III Biology, Medicine and the Future of Mankind

Dr Wallerstein and argued that the balance of world power is now already in the process of being transformed in favour of the Third World countries.

Finally, Dr Holland noted that many new technologies have the capacity to greatly reduce boring and repetitive work; but he also raised the spectre of widespread technological unemployment which might accompany their installation, and he apparently could not concretely envisage basic social changes capable of relieving the burdens of such unemployment. He did, however, mention the pressures for protectionism in certain industrialised countries which are faced with foreign competition based on technological innovation; but he did not consider which sort of countries such protectionism would conceivably be directed against.

Let us now turn once again to the more detailed summaries of the positions taken.

In his paper, *La maîtrise de la vie: pour quoi faire?* Dr Bruno Ribes gave an overview of certain aspects of current biological and medical research, and he then considered some of the short- and long-term perspectives in these fields. The paper is divided into three sections, the first of which focuses on the dynamics of mastering biological processes and considers some of the factors propelling modern science on towards greater control over life; the second is a critical reflection on what 'life' itself is; and the third then raises and examines the essential question of the purpose to which mankind's increasing scientific and technological abilities can and should be applied.

First of all, then, it is fairly obvious that there are urgent objective problems which necessitate a growing control over various aspects of life. Everyone nowadays is quite well aware, for example, that "industry pollutes and perturbs ecosystems and that we are going to have to give more and more attention to the effects which industry is having on the totality of life: not only on our environment, but certainly also on the strictly physiological conditionings of our contemporaries". Likewise, population growth and consequent deficiencies of food supplies urgently necessitate the development of biological knowledge which will be of use for agriculture and aquaculture. In today's world, and especially in the industrialised countries, the maintenance of individual health is facing severe problems; it is clear that "health is to a large extent jeopardised by our own conditionings [and] our styles of life, not to mention the stress inherent in urban and industrial life". At the national level, greater and greater expenditures have been and are being devoted to cope with worsening conditions of popular health. And, finally "it is clear that [all of] these same problems are being posed at the international level, given the fact that today we are witnessing an extremely rapid degeneration of the genetic patrimony of mankind, with a potential for hereditary illnesses, which, by the way, is increasing in proportion to our progress in mastering life". For example, because diabetics can now survive, their number has multiplied "by 2012 per cent in 28 years"; and because of breakthroughs in overcoming childhood afflictions "the augmentation of mental diseases ... is today very considerable: at the present moment most developed countries have about a 2 per cent rate of mentally handicapped people [in their populations]; it is probable that this rate will reach 8 to 10 per cent in the 2050s." Besides the stimulus provided by these problems related to industry, agriculture and human health, it is also "manifest that biological research is being stimulated by formidable interests which are eager to obtain results that will allow them to turn their current investments into profits and which influence the end-results of scholarly research".

These general remarks concerning the perspectives for biological research in the foreseeable future can now be illustrated by considering in turn three concrete directions in which this research "is undoubtedly going to be intensified in a most spectacular way in the years to come". The first of these directions lies in the exploration of the marvellous possibilities that have been opened up by the bacteriological 'industry' – i.e. exploration of the possibilities offered by making bacteria produce a certain number of desired substances, "whether medicines and vaccines (e.g. insulin) or products for current consumption (e.g. silk, which is not a product of the silkworm itself, but rather of bacteria which live in it). Perhaps soon enough it will be possible to produce methanising and de-methanising bacteria. The great dream is to be able to make nitrifying bacteria which will enable plants to fix nitrogen; but in that case it will then be imperatively necessary to produce de-nitrifying bacteria, since it will be necessary to respect the nitrogen cycle in the world. This last eventuality shows that we are being posed with a crucial question: we are capable of doing many things, but, all in all, our knowledge is extremely limited. . . ." It must especially be stressed that contemporary research "unavoidably takes place within a systematic perspective. However, if we do know a certain number of things quite exactly, our knowledge of totalities and of systems is very poor. From this point of view we are acting without a global perspective."

A second area of biological research in which great strides are undoubtedly going to be realised (although illusions here should be guarded against) is that "not of eugenics in the traditional sense, but of direct genetic control, by manipulation of the genes themselves for the purpose of overcoming certain hereditary diseases". This possibility confronts us with a major problem which is especially important for biologists: viz. they act in time, but the full effects of their interventions appear only gradually. In other words, if one knows the power of a particular bomb and sets the detonator, one can predict fairly certainly what results will occur within, say, thirty seconds. However, in dealing with biological processes, the chain of cause and effect is felt in the following generations; i.e. one is acting on long-range temporal processes and, moreover, "in a theoretically irreversible manner". On the other hand, biological intervention in time is now requiring more and more sophisticated forms of compensation: if one succeeds in provoking a particular mutation or alteration, for example, one is then immediately obliged to counterbalance this change in order to re-establish homeostasis. "We are thus becoming involved in a more and more rapid intervention-compensation process . . . [And] for modern medicine, death occurs

III Biology, Medicine and the Future of Mankind

when compensation is no longer possible. Now, our control over life — already at the industrial level, given the level of pollution, and probably at the agricultural level as well, given the uniformisation of plant cultures — will oblige us to put into effect more and more accelerated systems of compensation; and it is necessary to admit that in these circumstances we are unable to master the time factor."

A third area in which biological research will continue to come up with surprising breakthroughs is that of the modification of human behaviour. According to Dr Ribes, possibilities developed in this area raise problems that are much more severe than those connected either with bacteriological research or with genetic manipulation. "Without a doubt, we are going to be able to intervene deeply in the processes of the human psyche, and everything seems to indicate that most people will go along with this quite happily", because many of them are living in conditions of misery and stress. "Often without knowing it, many have already let themselves in for highly suspect forms of manipulation." The reader will readily recall certain types of lighting installed in supermarkets in order to psychologically dispose clients to buy more. Dr Ribes also recalled the case of certain schools in which tea-time has been given a novel twist: each student has his cup of tea, but these aren't just innocent refreshments. Students judged apathetic or lackadaisical have their stimulants, troublesome students have their sedatives; but "no-one has taken into account that apathy or troublesomeness, in fact, pertain much more to the psychological than to the physiological order".

The crucial question with which we are faced by the general development of biological knowledge is whether "under the pretext that we have certain powers, we [thereby] have the right to do everything and to let everything be done" that is able to be done. Should we "with ever more temerity exercise the powers that we have acquired over life"?

In answering this crucial practical question, one is led to consider a more theoretical one — viz. what is life? "All biologists look forward to the day when their knowledge will cover all of the structures and processes of a living organism. But even supposing that this goal is attained, we shall still not know what life is. For life is not only a given number of coded, programmed and regulated chemical substances which form a system"; it is also a dynamically constituted totality. Even "assuming that one day we might be capable of constructing a bacterium, we will not really know which dynamic has thereby been initiated or how, starting from such a bacterium, this new life will evolve".

In trying to grasp the essence of life, controversies about evolution constantly keep coming up; and "the problem of the evolution of life is at least as important as the problem of what constitutes it materially". But the debates on evolution are, of course, themselves quite thorny. Is evolution, as Jacques Monod claims, a series of chance events governed by necessity? Or is it, as Motoro Kimura believes, a series of coincidences triggered off by coincidences? On the other hand, do functions determine structures, or do structures determine functions? Without going into the details of such debates, it can be noted that many of the problems undoubtedly spring from our spontaneous representation of evolution "as a sort of rectilinear predetermined process generated by a directing principle. ... Actually, evolution must be conceived in a systematic way: provoked by a totality of actions, reactions and regulatory processes, within a four-tiered structure of living particles, organisms, populations and relations between the living being(s) and the environment. In this perspective, structures and functions are in close correlation, reciprocally engendering and reinforcing each other."

This said, two types of evolution should be distinguished. There is first of all what might be termed a 'horizontal' evolution, which moves towards improved functioning within a fixed horizon. "Most of the great biologists have nowadays shown the extent to which living beings are controlled by extremely rigorous structures and by regulatory processes — i.e. by a solid internal necessity. 'Horizontal' evolution refines and reinforces such necessities; it is closely bound up with them; and it works towards a perfecting in accordance with necessity...."

"But inherent in life there is also a 'vertical' evolution which pushes living things — it is not too well known how or why — to enter into qualitatively superior forms of existence...." Now, "among scholars, the debates [about evolution] usually evolve around the question of the perfecting of the individual" and concern horizontal evolution. "In fact, we are unable to understand how, given that the structures of the living being are so necessary and so strictly regulated, it is possible for evolution towards qualitatively superior forms of existence to occur." But the more that one considers this problem, the more it seems that the fundamental dynamic of this vertical evolution consists of "something extremely mysterious inscribed in the depths of the living being" — viz. "its openness towards another living being". This openness is "already manifest at the biomolecular level, since the molecular chain proper to living things is dissymmetrical, and thus able to hook up with other molecular systems". And it is also "more and more evident as one ascends on the ladder of life. Life is always in a state of profound disposition towards the other, constantly awaiting and stimulating the other. Openness towards another living being is inscribed as a structural and structurising law of every form of life." And this openness advances towards vertical evolution "by means of combination or co-penetration with another living being". Thus, it is because a virus penetrates a bacterium that the former is able to modify its genetic make-up. Now, such co-penetration has been constituted by life as a fundamental law, presiding [for example] over the apparition of sexualisation, which is not primarily a law of reproduction [since reproduction had existed long before], but a law of co-penetration, offering an enhanced possibility for the emergence of qualitatively superior forms of life.

"It is necessary here to meditate deeply on what sexualisation means. At the level of bacterial reproduction, the individual divides into two, on an average of once every two hours; there is multiplication or quantification of a single being, two from one. Once sexualisation has appeared, one passes from the order of quantity to that of quality; there is no longer a two from one, but a one from two. More importantly, the entire relational state between living beings becomes

III Biology, Medicine and the Future of Mankind

essential to life itself. No longer does the individual possess the key to the definition of life, if it is true that the definition of life (as opposed to matter) implies reproduction; the individual shares this definition of life with a partner. Henceforth life is [something common] between two living beings. The unity between two living beings, what sort of reality does it have?" Without being able to answer this question, one is incapable of grasping what life is. The preceding observations lead one to conclude that "all forms of research concerning the finality of life are vain. The position of the finalists is strictly untenable, and that of the anti-finalists is no less so. There is a search and an exploration which passes from living being to living being [and] which moves through and beyond us. Life consists of making the other desperately [à faire l'autre éperdument]. ... Desperately – i.e. with a great disdain for losses, without knowledge of where things are going to end up, but with the eventuality always present that a breakthrough might be produced into qualitatively higher forms of existence. Anyone who wants to define life must take into full consideration the 'logic' of this dynamic which traverses all living beings. The living being is the in-itself (i.e. obeying a profound necessity) which never ceases to ex-ist (to go out of itself) in the other. And the more we progress on the scale of living beings, the more we perceive not only that internal necessities become progressively more rigorous, but also that the urgency to ex-ist in the other becomes more and more essential."

According to Dr Ribes, it is imperative in the world today to assume the responsibility inherent in this logic of life and, if possible, "to add to life a supplement of 'logic'". In this light, several suggestions can be made in regard to perspectives for using the various forms of control over life which are being provided by biological research.

The first suggestion is based on the proposition that the purpose or finality of life consists of "desperately making the other". "There is nowadays a certain divergence among biologists: some of them are carrying out research and trying to understand the 'how' of life; others are concerning themselves immediately with the 'why', so as to assign a finality to life.... [Nevertheless] we are being compelled more and more to recognise – in regard to industry, agriculture and world health policies – that one cannot ... propose a 'why' without modulating it with a 'how'".

Secondly, in regard to the various forms of current research aimed at the modification of human behaviour as well as in regard to the various forms of genetic manipulation, the great concern must not only be that of pushing forward horizontal evolution, of determining life or even 'perfecting' it. While such projects are undoubtedly necessary, "our fundamental aim must be to de-fine life, to de-termine it and not to shut it up within ready-made moulds or frameworks. This is all the more necessary because life is and must remain adaptive", if it is not to be wiped out. "Quantities which we may today consider as being the most desirable are not necessarily those which will be the best in the world of tomorrow, given probable modifications in the environment as well as transformations in the social order."

Likewise, efforts in the fields of biology "must aim at preserving, restoring and intensifying the relational character of living beings". That is obvious at the theoretical level; but at the practical level, when one sees what is being done or being prepared (particularly in regard to genetic manipulation and modifications of human behaviour), one can justifiably ask to what extent the relational character of human life is going to be safeguarded in the future.

Concomitantly, Dr Ribes then insisted that "the singularity of the individual" be respected in biological research. In particular, "we do not have the right to say straightaway that the biological comes first and thus conditions the psychic; it is often the inverse which is operative – i.e. the psychic determines the biological". Thus, it will not do to intervene too facilely, at the biological level, "for in reality we will only be treating the effects and not the causes of the ailment or afflictions, and the causes are often to be sought within the psychic framework of each individual".

Finally, "it is important for us to preserve the historical character both of particular living beings and of humanity as a whole". According to Dr Ribes, "we are living at a moment of time in history, and it is not our place to level this history off or to normalise it. We do not have the right to prevent mankind – by planning health too quickly or by intervening biologically – from going through the adventure of an emergence, of becoming otherwise. In effect, this adventure of becoming the other is the only real meaning that can be prescribed to life. It is necessary for us to realise today that a biology which would aim at suppressing the other, and at suppressing the adventure of the other in us, even within our psyche, such a biology would be foolish and profoundly illogical."

In his paper, *Restructuring a framework for the assessment of science and technology as a driving power for social development: a bio-sociological approach*, Dr Mori Yuji extended the anthropological methods developed by Prof. Imanishi Kinji to contemporary power relationships; and he applied Imanishi's non-Darwinian theory of evolution in order to analyse such power relationships at the systems level.

Dr Mori began by remarking that although the social sciences themselves are devoted especially to investigating the unique characteristics of human society and its development, they have, nevertheless, been strongly influenced by the biological sciences. Since the time of Darwin, for example, the theory of biological evolution has acted as a major challenge to social ideas. At the same time, however, developments in biological theory can be shown to reflect the particular social circumstances in which they arise.

The historical merit of Darwin, of course, lay not in the discovery of evolution, but in his having "firmly established a scientific theory to explain the causes of evolution". Nevertheless, the Darwinian approach is marred by what might be called its fixation on the individual organism and its genes. "The quintessence of Darwinism is that: (1) an overproductivity of living beings exceeding the possible bounds of survival occurs; and (2) superior-inferior differences exist between individual organisms, and, hence, as a result of the

III Biology, Medicine and the Future of Mankind

'struggle for existence' between these organisms, only the fit are able to survive. In this process natural selection operates." In other words, the 'selection' of 'fit' individuals is left to 'Nature'. This position is common both to Darwinism and to neo-Darwinism; and 'natural selection' defined in this way has "come to be regarded as a major cause of evolution".

"... One of the unique characteristics of modern culture is the existence of the following [contradiction]: on the one hand, people have a high degree of trust in the axiomatic nature of natural laws; on the other, they have a vagueness and uncertainty regarding [the nature of] social laws", and there are few scientific laws explaining specifically social phenomena. In this regard, it is interesting to note that "Darwinism and neo-Darwinism have both been shaped by the influence of a competitive society; however, when evolution [or development] based on competition and [the principle of] the survival of the fittest were established as natural laws, they were also, in fact, accepted as laws governing society. Needless to say, nowadays such ideas have so fully penetrated people's life that they are regarded as common-sense. Moreover, an extreme form of Darwinism, social-Darwinism, is now being emphasised; and ideologies of big-powerism, war and aggression applicable to human society are being marched out." All in the name of the progress of the human species.

Mention of social-Darwinism cannot fail to raise the subject of aggression; and, in this connection, one might recall that in 1963 Konrad Lorenz sparked off a lively debate with his book *On Aggression*. Lorenz detailed how "both animals and humans are equally endowed with aggressiveness". For animals, however, "aggressive behaviour does not lead to the defeat of the other power but rather becomes a bond of solidarity between the animals"; and thus "aggressive behaviour functions to maintain order in the animal world". With human beings, on the other hand, the manifestation of aggression may lead to killing the other party and may lead to war. "The beginning of the 1960s was the time when the ... cold war reached its extreme and the United States' invasion of Vietnam became 'America's Vietnam war'. The formation of Lorenz' theory of aggression and the debates around it could not have possibly occurred in isolation from this historical setting." Yet, without entering fully into the details of this debate, it can be observed that "Lorenz formulated and developed his theory by focusing on the aggressive behaviour of [individual] biological organisms and human beings. In this respect, the core of his theoretical construct is identical with Darwin's".

Another theory based on neo-Darwinism and attempting to analyse human nature and society is socio-biology – a subject which has received considerable publicity since the publication in 1975 of Edward Wilson's *Socio-biology: A New Synthesis*. "Socio-biology can be defined as an interdisciplinary science drawing on biology (particularly ecology and physiology), psychology and other social sciences. Research covering such fields is also referred to as bio-sociology and animal sociology." Leaving aside Wilson's own motivation for his research work, it is clear that the well-publicised debate concerning this scientific theory reflected social tendencies particular to the period of transition from the 1970s to the 1980s. As noted above, "behind the debate on Lorenz' theory of aggression was the manifest display of human aggression in the Vietnam war.... The identical situation does not exist at present; in fact, it might seem as if behaviour directly exhibiting aggression has already been hidden from view. Still, oppression has not disappeared, nor has opposition to oppression; it seems, rather, as if a powerful, complex, oppressive organisation — one that it is difficult to come to grips with — is gradually blanketing the world. Isn't this the reason why the path to liberation is no longer clearly visible?" And it is in the light of the "political, social, cultural and natural crisis of the modern world" that socio-biology is advanced as an attempt at finding a solution to the problems faced by mankind.

How does socio-biology approach these problems, however? If one consults socio-biological literature, it is apparent that at the centre of discussion is the interaction between mind, action and genetic structure. Positions on the different sorts of interaction between these three — e.g. in human beings versus animals — may allow one to throw light on the distinction between instincts and the ability to learn. Nevertheless, like other forms of neo-Darwinism, socio-biological theory is based on the individual organism; and for this reason its statements about society remain open to objection. "Simply adding together human beings and their actions does not make a society. This is the problem in the socio-biological approach to human society."

Now, although any particular science must begin with examining the elements composing a system, at some point or another it is necessary to examine the functioning of the system as a whole. And "the bio-sociology of Imanishi Kinji is an example of a non-Darwinian theory of evolution that approaches things from the systems level. Imanishi adopts a holistic point of view". "In contrast to Darwin, who based his theory on the overproductivity of living beings and variations between organisms (i.e. the construction of a theoretical system based on the organism), Imanishi takes into consideration the historical and social nature of a species to construct a theoretical system based not on an organism belonging to a species..., but on a society of species.... That is why his theoretical system is called bio-sociology." For Imanishi, every biological species constitutes a society; and "society is a universal phenomenon", of which human society is only one example. Interspecies sociology and intraspecies sociology constitute the two main divisions of bio-sociology; "the former takes into consideration geographical and historical factors; the latter ... includes the level of the individual organism and of a society of species".

In contrast to the Darwinian tradition, "Imanishi's theory of evolution gives primacy not to genetic mutations within the individual organism, but rather to the way in which an entire species is transformed". Likewise, according to neo-Darwinism, mutation occurs randomly via natural selection; according to Imanishi, however, "mutation is incorporated into a species systematically". Dr Mori stated the major conclusion of this bio-sociological theory as follows:

III Biology, Medicine and the Future of Mankind

"congeneric organisms (which are the same morphologically and functionally, or systematically and behaviourally) must undergo change in the same way when the time for change arrives. Moreover, as a result of these changes brought about by mutation occurring in all the organisms of the species, mutation improves the species' fitness to win." According to Imanishi, the development of human society followed this path.

The formation of herds is a characteristic of higher animals; and "Imanishi stresses the importance of sister relations as the origin of group life among the higher animals. In line with this, it is possible to trace the origin of culture as far back as the establishment of the herd. It was through farming that human beings came to take a completely different evolutionary path from animals. What is of importance here is that farming allowed the production of a social surplus. In contrast to biological evolution, therefore, which occurred by a metamorphosis in the form of the body itself so as to adapt to Nature, human beings evolved as the result of modifying the environment and achieving independence from Nature. To put it another way, human beings evolved through culture. In fact, we can consider that speciation in humans came about through culture and that culture is the same as species in the animal kingdom. What was most important in bringing about [the change from animal to human] was the use of tools. This became technology and is at present referred to as 'science and technology'. The production of social surplus by the use of technological power brought about the stratification of human society, owing to the unequal possession of the surplus. Between societies, too, it is creating a hierarchical structure. As in the animal kingdom's food chain, groups within [human] societies or societies [States] themselves are appearing as predators. In this sense, human beings today are living in a dual hierarchical structure."

Within the bio-sociological framework of analysis, "three levels of production and consumption can be distinguished: (1) the organism level; (2) the social level; (3) the political level". In Imanishi's theory, of course, the social level is characteristic of all biological species. The political level, on the other hand, is encountered only in certain human cultures; and "the origin of the political level can be found in the attempts to solve the problem of the possession and distribution of the social surplus that appeared as the result of human beings starting an agricultural life.... [And] the stratification of species society was brought about by political power." Within human societies the social and political levels "generally overlap"; but there is today a great gap between the two which is perhaps exemplified most strikingly by the modern nuclear weapons systems whose "sole function is to play a political role" and which are a "factor impeding social development".

Another example of the current gap between the social and the political levels is seen in regard to the reproduction of the human species, especially with the population explosion in the developing countries, although this is often explained as simply a biological phenomenon. According to Darwin's theory, for example, "the overproductivity of living beings beyond the survival limit becomes the cause of the struggle (competition) for survival and promotes evolution". Thus, it is assumed that an invisible hand is working for progress, and that things are going as best they should. Nevertheless, one cannot fail to remark that the countries affected by population pressure are those which (1) "have been controlled, exploited or oppressed by international predators"; (2) are "rapidly attempting to accumulate social surplus through an internal ruling class of predators"; or (3) are beset by both internal and external predators. According to Imanishi's theory, physical needs in human society are always mediated by culture, and human survival must be guaranteed through culture. In the case of population pressures, "it is not because of overpopulation that poverty and starvation occur"; but, rather, it is more and more difficult to accommodate a growing population when exploitation and oppression are creating poverty and starvation. And "so long as the structure of robbery, exploitation and oppression remain the same", various types of 'aid' will "merely aggravate the situation".

It was noted above that "what gives production activities in human society their unique characteristic is production based on science and technology", and "one can call this culture". It will also be recalled, however, that in the process of evolution there emerged a variety of human cultures which in certain ways resembles the variety of animal species. Yet, whereas different species remain distinct because cross-breeding does not occur, "culture comes into existence through mutual influence and receptivity"; culture tends, in other words, towards universality.

Modern science and technology have "freed production power and brought about social development"; but it cannot be overlooked that they were "perfected in the European world because of the industrial revolution", and thus "their natures were formed against a European social and cultural background". Hence, it is possible to speak of "the cultural element in science and technology". And it is "quite natural that in the European world both [science and technology] have been a consistent whole rooted in society. At the same time, however, it is natural for problems to arise when Western science and technology are transplanted into a society with a different culture". In order for "the universal and general principles of science and technology" to contribute to global social well-being, it is therefore necessary to define and meet the objective needs of the non-European cultural areas. This is "indispensable in order for science and technology to root themselves ... as culture" in these societies.

In conclusion, Dr Mori said that "modern science and technology have brought into play enormous, ... productive powers and have made possible rapid communication and the high-speed transportation of goods and people". But he wondered whether the very achievements of modern science had not given rise to the — necessarily temporary — illusion that human societies can be manipulated in the same way as things can be. In this regard, one can distinguish two contradictory approaches to social development. "The first is to strengthen social organisation and social order. The second, in contrast, is to increase the amount of freedom — i.e. to expand possibilities. We can say, for example, that modern science and technology have brought about an increase in productivity through the high rate of organisation of a production system; on the other hand, however, they have also made the people who work into no more than parts of the production machinery. If this is the case, then modern science and technology are working for the oppression of human beings. So long as an increase in material production and a decrease in work do not tie in to the construction of a social system that increases social and human freedom, it is clear that modern science and technology in their present forms lack the power for [contributing to] social development."

Dr Milanović next presented a position paper entitled *Human aspects of the medical sciences: medical technology and the physician's responsibility*. The physician's responsibility is a very delicate matter and cannot rightly be limited to his legally defined duties. If the physician only complies with accepted norms, but does not do all that he should in caring for his patients, he will certainly, for example, feel a burden on his conscience. The contemporary physician is "torn between rigorous regulations, exceptionally high expectations from society, his own human needs, his other obligations and tasks and his own conscience"; according to Dr Milanović, he is "in the most complex existential situation, burdened with dilemmas by which no other profession is attended". Let us, then, consider some of the specific problems faced by today's medical practitioner.

A doctor first of all encounters people, his patients, who find themselves in a position or role for which they have not been prepared. The physician can, of course, only effect a cure if there is at least a minimum of co-operation on the part of the patient and his environment; and although it is therefore not legally the physician's responsibility, he is, nevertheless, concerned with the total state of the surroundings and with the existence of cultures that "are in conflict with modern medicine" and human health.

Another specificity of the physician's role derives from "the exceptionally high level of social expectations" which he must fulful. "If the public detects a lack of sympathy with those in affliction, [or if it detects] bureaucracy and formalism in medical organisation, it reacts very severely. It would certainly be wrong to advise ... the same patience in waiting for urgent medical aid as that shown [towards] the slow rhythm of work in today's administration"; but it is, nevertheless, all too obvious that physicians occupy a very difficult and contradictory position in any society in which 'money makes the world go around'. The disparity between serving the needs of others and lining one's own pocket "will inevitably be expressed in the controversial functioning of medical institutions and the behaviour of the physician". Apart from such 'external distorting influences', however, "the sphere of medicine is an area of creation", in whose "humane character ... lies the intention of searching for possibilities of exceeding ourselves and the situations that limit us, for higher forms of unity with our own nature and with other people".

Its own body of special skills and techniques is, of course, another aspect of the specificity of the medical profession. As in other fields, this medical corpus is now too large to be mastered by any one individual. "The constant development of investigations and the application of their results in diagnosis and therapy have brought about continuous specialisation", involving, on the one hand, the constant narrowing of specialists' skills, but, on the other, more and more penetration, exactitude and efficiency. This constant specialisation, however, has resulted in a "quite unpersonal" relationship between physician and patient, more like that between things than between living people. "The present-day patient . . . in many respects reminds us of production material travelling on the assembly line"; rarely is a diagnosis made nowadays on the basis of personal contact, knowledge and sympathy. As the economic concerns of large institutions require rational expenditure of time, energy, materials and the greatest possible speed at work, the patient's files may become thicker and thicker, but there is no one person who knows very much about him; and bureaucratic orientations in organisation tend to increase the 'collective irresponsibility' of individually responsible experts.

What alternative is there to this situation? The medical profession must, obviously, continue to use modern techniques/technologies, "without which there is no efficient medicine": it must, likewise, continue to observe economic principles. An alternative, however, will require medicine "to return the human character to the relationship between the physician and the patient. Not only biological, chemical and other techniques, but also the human word, a human attitude and understanding on the part of the physician, must again become means by which health is restored."

Another aspect of the social responsibility of the physician is connected to present inequalities in the distribution of and access to medical facilities. For example, in all advanced countries, "regardless of . . . whether medicine is socialised or practised privately", patients seek access to first-class medicine in serious and complicated cases; and sociological investigations show that patients with more favourable social status (higher education, income, reputation, political power, etc.) tend to frequent the more reputable medical institutions. Although this fact highlights the extent to which the distribution of resources constitutes "a limiting factor . . . [for] both the professional possibilities and the responsibility of the physician, the way out lies certainly not in the administrative manipulation of patients, but in the general (scientific, technical, technological, personnel) advancement of all medical institutions to a high level corresponding to the standards of modern science". Doctors can and must display initiative in working for such a solution, and in this project they "will certainly be supported by all democratic forces of the society".

As a social institution, medicine is very much imbued with an ethic of high individual professionalism — i.e. with "an ideal of the expert who acquires a good reputation through irreproachable work". Although a holdover from the times when medicine was a trade, this concept "contains in itself very important elements without which one cannot build the internal feeling of responsibility ... towards the patients. This concept implies constant observation, acceptance and application of the results of science and technology; and any instance of lagging behind modern developments, and thus of ignorance and inefficiency, it proclaims unethical. Although this ethic takes for its norms the exact principles of science, it represents a very important accelerator, not only of scientific progress, but also of the humanisation of medicine. . . .

"Without [however] underestimating the professional concept of responsibility of the physician and the role which it has played and will continue to play in the development of medical science and organisation, it seems . . . necessary to emphasise that [this conception] is inadequate for the needs of modern science and of progress." The individual scope of the practitioner's responsibility to himself or to his profession does not adequately reflect the realistic demands made by modern society. The concept of medical responsibility should thus be redefined as "responsibility to the community – [and] not only to the local or national community . . . ; thus conceived, the principle of responsibility to the community includes, first of all, the responsibility of the physician to himself as a human being who seeks his affirmation and his happiness in the accomplishment of the health and happiness of other people, of society as a whole. Thus conceived responsibility does not permit reduction of the physician's responsibility to the professional and routine performance of his duties, to the separation of his personal from his social life, but rather expands it into a creative, critical attitude towards oneself and others, towards the social totality as a whole."

It can be hoped that this redefinition of medical responsibility will guide physicians to a better understanding of the tasks and possibilities facing them today; but the place of medicine in social life should not be idealised. By itself medicine cannot overthrow antiquated forms of social organisation; "neither can it become a starting point for deeper social transformation", for it is always only a part of a given society. But, nevertheless, "in medicine, more than in any other area, as through a prism there are refracted and resolved the problems of the human individual, who is equally strangled by the depredations which time brings on and crushed by the rushing wheel of modern civilisation. With the necessary ability to appreciate the pains, desires, anxieties, needs and capabilities of modern man, the physician can become an important factor in the solution of [modern man's] existential problems. . . . And this openness and human determinedness of the physician, his critical engagement in the struggle to change the world to which he belongs, his entrance upon the broad social stage, and his increased load of responsibility can be equally useful, both to mankind and to medical science."

Discussion

Dr Pandeya began the discussion by sharply opposing Dr Mori's attempt to view contemporary human problems in terms of a general bio-sociological, evolutionary process. According to Dr Pandeya, arguments of this sort are based on "analogical extension" and only serve "to disguise issues and to mask uncomfortable ... exploitative social realities". Such an approach becomes even more dangerous when framed in terms of sophisticated systems analysis; and if systems analysis is to be used at all, one would do better to stick with the framework put forward by Hegel — viz. that of the master and the slave, which "has perfect functionality, interdependence, mutuality, stability and equilibrium" and which in its simplicity meets all of the requirements posed by more modern analysts.

Dr Pandeya characterised the bio-sociological approach as a "mixture of social-Darwinism and systems theory", to which two basic objections can be made. First of all, "presenting problems of social conflict, of exploitation and of deprivation in this large macro-context ... amounts to looking at our problems from the farthest planet away from Earth, where all of the significant, important, concrete, historical and social details are obliterated" — so that what is left is a very harmonious, abstract outline. Seen in this way, "what was actually primitive accumulation dripping with blood becomes a very beautiful harmonious process, which has conferred on mankind nothing but positive benefits". The tremendous industrialisation, the great increase in productivity and the scientific-technological "appropriation" serving these goals are portrayed simply as a uniform sort of progress, although the fact is that the greater part of humanity has had to pay very dearly for what has occurred. One may be able to demonstrate the existence of certain patterns at the ecological and subhuman levels; but the method of extending these patterns by analogy in order to transform social and political realities in the contemporary world does not make sense and is "dangerous". This sort of approach is one typical of "those notorious characters whom Karl Marx once described as 'the great harmonisers'". Such people nowadays "sing the song of interdependence and of global unity", and their effective social and political message is that mankind is at present living in the best possible world that could ever be achieved. Those who seek to transform the world "must stand forewarned against such harmonisers, however sophisticated, pleasing and comforting their 'systems-frame' and their 'long-range macro-perspective' may be".

In regard to Dr Milanović's presentation concerning the responsibility of the physician, Dr Pandeya said that the problem of improving health is not merely a problem of professional medical ethics, or of improving the access of certain social groups to high-level medical institutions. The major parties responsible for the present state of medical practice are neither individual physicians nor national medical institutions, but the transnational corporations which control the manufacture and distribution of medical equipment and pharmaceutical products. Individual physicians are often under the illusion that they are simply performing necessary social functions according to an admirable professional code; but, in fact, the medical and pharmaceutical transnationals are primarily concerned with converting problems of human health and disease into a "flourishing profit-making industry". Such corporations have a return of "more

III Biology, Medicine and the Future of Mankind

than 400 per cent... on a minimum capital investment", which means that they have one of the highest profit margins in the history of capitalism. Their hold on Third World countries is especially strong and harmful.

In the second intervention of the afternoon, *Dr Celso Furtado* began by saying that it is essential for any discussion about technology to make an explicit link between human actions involving given techniques and the desired goals of such actions. According to Dr Furtado, the conference had so far witnessed some confusion, because it had not yet been sufficiently stressed that it is power that constitutes the link between various forms of technology and the general goal of the transformation of the world. "By definition technology generates power, because efficiency in action is power. To mobilise or activate resources in order to increase efficiency is to generate power. Thus, the control of technology is a power resource." In order to discuss the transformation of the world realistically, therefore, it is necessary to consider the instruments of power as power resources; but it is impossible to get to the heart of the matter if one limits oneself simply to discussing tools as such.

On the question of technology forecasting, we have to remember that technology is not something that is being generated 'on another planet'. "To forecast technology is by definition to postulate the goals that people... are striving to reach", and various technological forecasts, therefore, necessarily assume that the world of tomorrow will have particular characteristics.

On the other hand, "what is the fundamental element that is commanding technological progress, now? Everybody knows that it is the arms race. Half of the resources used to develop technology after World War II came from the military budgets. The world that we are building now by means of technological progress is a by-product of the arms race." Without keeping this fact in mind, one cannot go far in understanding the present transformation of the world.

As a final point, Dr Furtado spoke about the significance of certain trends in the world economy. Although there was a long-range historical trend during the past few centuries for income to concentrate in a way advantageous to certain regions, Dr Furtado doubted whether any practical conclusions for the present moment could be drawn from this historical observation. Not only are the data available for the original trend very limited, but also it is very difficult to compare different phases in history. On the other hand, "from 1948 up to 1974, [there was] a rate of productivity increase of 3.5 per cent per annum in the developed capitalist countries and of 2.5 per cent in the Third World. If we take into account the increase in population (which was much larger in the Third World than in the centre of the capitalist system)", it can be argued that there was "a higher rate of increase of production, though not of productivity, in the periphery than in the centre of the capitalist world". Thus, likewise, the global market has been expanding more on the periphery than in the centre. Moreover, the main increases in the centre were in the service sector; and "the accumulation at the level of productive forces linked to the production of tangible goods was larger in the periphery than in the centre".

Expanding on his intervention on the previous day, British Labour MP *Dr Stuart Holland* stated that he did not consider the potential negative effects of technical progress to be "unavoidable and irreversible"; but not to subject techniques and technical progress to social control constitutes "a major failure on behalf of societies and political systems". Even when social control is exercised, however, "there is not simply a finite solution to many technical problems — because these techniques themselves throw up new options, new problems . . ." and because "we have to admit that there are different options for social control in different societies with different concepts of development".

Nonetheless, "the negative consequences of many of the new forms of technical progress have still been underestimated, especially by governments". For instance, the Central Policy Review Staff unit attached to the British Government for the last ten years has "predicted that if we applied the available new technologies, the robotic and semi-robotic technologies, to the production of goods in Britain, then within 30 years we could produce all the material goods which we would then need with only 10 per cent of the existing labour force. In other words, a 90 per cent technological unemployment.

"To give one illustration, in some countries today on a purely technical basis, [it is possible to] produce steel in modern, shore-based, integrated plants with a quarter of the labour force employed in some European countries. However, this productivity is not simply a net gain. The obverse of the technical capacity is an imbalance between the steel industries of economies such as Japan or South Korea, and the decimation of virtually entire industries in some of the developed capitalist countries. And this is leading to very significant pressures for protectionism. . . . What this implies is that protection against technical progress and competition from very low-wage economies elsewhere in the world may well give rise to various forms of nationalism. But to stop that argument there is far too simple, because the issue is not simply [one of] free trade versus nationalism. It cannot seriously be argued that countries and societies should allow new technologies to be applied whatever the social consequences. . . . Many of us know well the force and dynamic of political and economic power which lies behind an unequal competition between capital and States . . . [and] the real world of technical progress is not the harmony and welfare of the competitive model and its myths, but a new international economic disorder. . . ."

Anyone seeking to transform this disorder, however, must realise that it is not enough to think in terms of a "simple model of transformation". Such people must also realise that "the applications of techniques and the adaptation of systems has to be multiple; it has to vary, it has to allow a very high degree of relative autonomy to different groups, regions, States and movements". On the other hand, in more practical terms it should be kept in mind that "many of the new techniques are not highly energy-absorbing". And they are also liberating, because they can do away with much of the boring, repetitive work which alienates many people for most of their working hours. This, in turn, requires one to consider the "redistribution of the massive new productivity made poss-

III Biology, Medicine and the Future of Mankind

ible by these technologies" as well as "the redefinition of the world of labour in different kinds of societies, and therefore [the redefinition of] the decision-making process".

Concerning international relations, Dr Holland said that some of the main problems being posed at the end of this century and the beginning of the next, such as armaments, raw materials and energy, "need a more international framework; but that may have to be based on less internationalism in all other respects. For instance, effective international security in armaments and avoidance of nuclear proliferation among many new countries only makes sense either if those countries are externally dominated (a solution which they rightly fight to reject) or [if they] have a degree of indigenous security in the management of their own economic and social base – i.e. can avoid internal and domestic crises without recourse to military adventurism. But to give greater economic security to such countries (or allow them to achieve it for themselves) may well mean that we have to accept greater autonomy or independence in their international economic policy, which could well mean their opting for less 'free trade' as well as resisting or rejecting many of the new generation technologies." Later, Dr Holland added that since European markets were now being subjected more and more to protectionism (not only in food, steel and textiles, but also in other areas), "we should not be surprised at counter-protection elsewhere".

In the early twenty-first century "many people will not have jobs in the way we now think of employment in the productive or 'productivist' sense. But to redistribute available and new jobs will mean relating social needs and services to the productivity increases which are possible from the application of technology in the productive sector. This will have to be [based] on different models and [implemented] in different ways in different parts of the world."

When Dr Holland had finished his intervention, *Prof. Tsurumi Kazuko* took the floor to make two brief comments on what Dr Pandeya had said at the beginning of the discussion.

Prof. Tsurumi first of all challenged Dr Pandeya's interpretation of Imanishi Kinji's system of bio-sociology. Although Imanishi had, indeed, sought to provide an alternative to the confrontational method of transformation implicit in the Darwinian model of evolution, according to Prof. Tsurumi, his approach was aimed not at harmonising irreconcilable forces in the social and political sphere, but rather at developing a non-confrontational method of transformation.

Secondly, in relation to world health problems, the Japanese experience is in accord with the position that "the multinationals are polluting the world and causing health troubles". Citing the example of Minamata, Prof. Tsurumi said that the different reactions to organic mercury pollution and its social consequences had demonstrated that "there are different types of doctors and of medical technologies". Some doctors directly or indirectly support the multinationals and thus delay the solution to health problems; others put their patients' interests first.

In the next intervention, Dr Rakić stressed the necessity of establishing "better links between the bio-medical, psychological, social and political sciences". The relationship between the individual and his changing environment is all too often overlooked, because, on the one hand, the biological and medical sciences concentrate almost exclusive attention on the individual, while the social and political sciences, on the other hand, fail to take him sufficiently into account. Nevertheless, it is clear that many ailments typical of today's society are the expression of difficulties in the dynamic between individual human beings and their environment. For example, at the simplest level, of course, there are adaptations and illnesses which result from increased levels of chemical pollution. In fact, it is fairly easy for the human body to produce enzymes which will allow it to cope with such pollution; but even so, the problems of long-term cumulative reactions must be taken seriously. On the other hand, it is much more difficult for the human body to produce enzymes which can counter the effects of what is now being called psychological or sociological pollution. The social and mental demands that are being made, not just on a few but on huge numbers of people today, are so difficult to cope with that they cause serious physical breakdowns in many of these people. Among the illnesses which result, one can number many respiratory diseases, digestive diseases, cardiovascular diseases and a whole series of addictions. The incidence of such complaints is quite high, but it is well known that bio-medical treatment alone can in most cases provide only superficial or short-term results. Such complaints are really socio-medical problems, and they require a complex multidisciplinary approach which will recognise and respect the dynamic interaction between individual human beings and the changing world in which they live.

IV

The Control of Space and Power

This fourth session focused mainly on control over geocultural space and geopolitical power, rather than over energy resources and the interplanetary regions. The key theme which underlay this session, negatively determining it and, indeed, placing it at the heart of the entire conference, was that of hegemony – the predominant control exercised by one or more foreign powers over the principal forms of the social life of a nation. The reader will have noticed that during the previous sessions criticism was repeatedly levelled against Eurocentric conceptions of the world; Drs Lefebvre, Štambuk and Mori, in particular, each formulated important objections from their own points of view. These objections were later to be further developed by several detailed expositions in the fifth session. Now, Eurocentrism can perhaps best be viewed as a particularly acute articulation of hegemony in the ideological sphere, and its force can be gauged by the extent to which it even penetrates mentalities about such supposedly 'objective' subjects as science and technology. Criticism of Eurocentric notions is undoubtedly an essential part of the struggle against hegemonic relations in the world today. However, a part should not be taken for the whole. Quite apart from the fact that many peoples within the European cultural area itself continue to be held in a state of dependence and poverty, how illusory it would be to think that the inequalities in the world are simply the results of narrow prejudices, misconceptions and ungrounded ideas. These inequalities are rather grounded in and embody a system of power relations, and hegemonic power lies at the heart of this system.

The struggle against hegemonism is the struggle of peoples to determine their own future within their own national boundaries. This struggle undoubtedly constitutes the principal task of subordinated peoples; and, in order to be effective, struggles to overcome particular inequalities – such as those related to scientific potentials and technological resources – must be integrated into this general one. Perhaps this last proposition can be understood as a specification of the general principle put forth in the first session which stated that science and technology can contribute to human liberation only if integrated into struggles for democracy. In any case, as stated by Dr Abdel-Malek in the last intervention here, the major problem today facing the nations of the Third World is that of maintaining their political and cultural sovereignty. As noted by Dr el-Kholy in his opening paper, solutions to this problem hinge on the ability of these nations to generate social and political systems which will ensure the efficient utilisation of the available human and natural resources. And Dr Pandeya pointed out that success in the fields of science and technology would require the formation of a broad, popular scientific culture.

The problem of the roles of science and technology in the contemporary world is nowadays far from peripheral to the general problem of hegemonism; and many of the participants in this conference stressed what Dr Vidaković called the mystification of their objective social functions. Dr el-Kholy, for example, noted that innovations imposed on the authority of an external power are more likely to serve as means of increased subjugation and alienation, rather than as tokens of some trans-historical progress; and he mentioned that 'big science', despite its economic advantages in the developed countries, is not necessarily cost-effective in Third World conditions and may contribute to intensification of the international division of labour if uncritically imported. Dr Silva Michelena, in turn, observed that technological optimism is an essential part of the developmentalist advertising being pushed by the transnational corporations to assure 'developing' countries of their bright prospects within the capitalist system. Examining the significance of nuclear energy for the countries of the Third World, Dr Pinguelli Rosa illustrated some of the typical complications that arise when a heavy technology is treated as an object of prestige, rather than as an instrument for meeting popular needs; but he also noted that these countries can ignore such technologies only at the risk of perpetuating foreign domination, and he stressed the importance of building up national independence in an all-round way. Dr Vidaković himself finally considered how, despite the various forms of scientific-technological optimism, the militarisation of the contemporary world economy is dominating the development of science and technology, harnessing them more and more to purposes of repression and destruction and thus obstructing the realisation of their great potentials for improving the lot of the peoples of the world.

Taking up this last point from a slightly different angle, Dr el-Kholy had remarked that the scientific and technological breakthroughs most needed, especially by peoples in the Third World, do not seem to interest the developed countries commercially; and, hence, the complementary strategies of national and collective self-reliance present themselves as practical necessities for the developing countries. On the other hand, however, he pointed out that, because the problems facing today's world are global in nature, adequate solutions to them will require appreciation of the contributions of all civilisations; and strategies of self-reliance are thus to be commended on theoretical grounds (related, inter alia, to the sociology of knowledge) as well as on practical ones. But the political prospects for bringing about such an appreciation in the 1980s — a sort of maximal programme, as it were — seemed admittedly rather bleak to Dr el-Kholy; and he therefore stressed the importance of elaborating a long-term vision of protracted transformation, on the basis of which a nation or group of nations would be guided in formulating and carrying out several distinct and

IV The Control of Space and Power

realistic tasks corresponding to a series of objective stages in the process of change.

For Dr Silva Michelena, progress in solving global problems is reducible to the evolution of the balance of power between the two main blocs and especially between the two superpowers. For him, the struggle against hegemony is therefore more or less equivalent to that against domination by the USA; and political prospects in the world consequently appear relatively bright for the decade to come. Approximating one of the positions in the debate within the non-aligned movement about whether the Soviet Union constitutes a 'natural ally' to the peoples of the Third World, Dr Silva Michelena stressed logistic support by the USSR as the most significant element in world politics today. One might ask about what happened to it during the revolutions in Iran and in Zimbabwe.

A number of consequences followed from Dr Silva Michelena's failure to examine the complex problems related to Soviet hegemony. He predictably gave little notice to the crucial importance of the principle of the self-determination of peoples and to the historical forces which make this principle operative. He was unable to envisage the real implications of contradictions opposing the lines of various nation-States and of national liberation movements to that of the USSR – hence his reading of the struggle in Eritrea, for example. He likewise failed to appreciate the significant contributions to world peace and liberation made by the non-aligned movement, that broad cross-ideological current which includes, inter alia, key socialist countries such as North Korea and Yugoslavia, etc.

According to Dr Silva Michelena, the West has a primarily economic interest in preventing the expansion of the Eastern bloc, which, on the contrary, has a primarily political interest in expanding. One can, of course, remark that politics itself is a concentrated form of economics; but more importantly it must be asked to what extent the expansion of the Soviet bloc differs from the extension of a hegemonistic sphere of influence. Certainly, there are few people who would deny the importance of the USSR's growing logistical capacity in the global situation today, but there are certainly also many who would differ with Dr Silva Michelena in regard to its significance. In this respect, it is perhaps pertinent to keep in mind a position put forward not long ago by Roumanian President Nicolai Ceaucescu, who, warning against attempts to re-divide the world into spheres of interest, noted that since 1975 the USA utilises principally economic means to expand its influence, while others – with inferior economic levels – make use of military force in order to reach the same ends.

All of which reminds one of the famous dictum by Clausewitz that war is politics carried on by other means. What is perhaps more evident nowadays than it was in the beginning of the nineteenth century, however, is the extent to which war is also the pursuance of particular economic goals – by means no longer alien to the 'normal' conditions of the economic process in either of the two major blocs. One of the chief contributions of Dr Vidaković's exposition lies in his having shown that the militarisation of the economy and the closely related perversion of the global development of science and technology are to be explained at the economic level by the fact that it is precisely in the militarised sectors of production that the highest rate of profit is nowadays to be realised. The struggle to secure maximum profits thus finds its necessary expression in the growing repressiveness of social structures, both in metropolitan and in subordinate societies.

Undoubtedly, the arms race and the menace of war lurking behind it number among the most overt and diabolical aspects of an entire repressive system, and they are linchpins, in particular, for holding together the system of hegemonistic relations at the international level. They are, therefore, being opposed more and more resolutely by peoples around the world, and especially by those in underdeveloped and dependent parts of the globe. Such peoples know well that, as pointed out by Dr Pinguelli Rosa, the heightening risks of holocaust (as well as the chief obstacles to their emancipation) are primarily the responsibilities of those who are so avidly going about accumulating the most incredible stocks of sophisticated weaponry — nuclear, of course, but also chemical, biological, electronic, etc. — while patting themselves on the back for signing documents designating the rate at which their arsenals will increase.

Is there anything that Third World countries can actually do about all of this? Can they mobilise forces strong enough to alter the international situation significantly and to give themselves some breathing space? Numerous examples could be adduced to support the claim that, given certain conditions, they can. To mention only two, the experience of the non-aligned movement in defusing potential confrontations between the superpowers would certainly seem to indicate the affirmative; and, at another level, the experience of the five 'Front-Line States' in support of the liberation of Zimbabwe is very positive in this regard. In general, the crucial point in answering this question is to be found in one's evaluation of the significance and complex inter-relation of the various forms of power; and during the discussion to this session two lines of argumentation emerged. The first, exemplified in Dr Furtado's intervention, stressed that since the end of World War II the economic potentials of the Third World had been significantly strengthened when compared with world levels; the inference was that countries in the Third World are therefore objectively more capable of building up their technological infrastructures. The second position, put forward by Drs Issa and Rasheeduddin Khan, for example, emphasised that those Third World countries which had not undergone a fundamental political transformation freeing them from foreign domination have typically less and less power to dispose of their own national resources as time goes on. Although presented in a somewhat antithetical way, the two positions do not necessarily exclude one another unless, of course, the need for political emancipation is denied. Taken together, they attest to the growing possibility and urgency of Third World nations' asserting their genuine independence, singly and collectively.

However, while most discussants agreed in principle with the strategies of collective and regional self-reliance, many were quite concerned with the various divisions within the Third World. Overcoming such divisions and achieving a unity

IV The Control of Space and Power

of action while accommodating differences are, of course, extremely difficult practical problems in the building of any united front; and there may, indeed, even be a few gaps which cannot be bridged. Such difficulties, however, cannot justifiably be allowed to serve as an excuse for the *a priori* dismissal of the possibility of united action among the peoples of the Third World. Were there no contradictions between the various participants in the anti-Fascist front during World War II? Who could be naïve enough to think so? Their common actions, however, were based not on an undifferentiated 'identity' of interests, but to one degree or another on a certain community of interests which had to be elucidated by protracted efforts and struggles. Perhaps this point is still valid today in the struggles versus hegemony and the menace of war. And a critique of the contemporary development of science and technology cannot be isolated from such struggles.

The first position paper presented to the fourth session was that by Dr Osama el-Kholy, entitled *Towards a clearer definition of the role of science and technology in transformation*. Dr el-Kholy noted that more than two years of intensive and world-wide discussions had been devoted to preparation for the United Nations Conference on Science and Technology for Development (UNCSTD), which was held in Vienna in August 1979. Since "practically all the important issues relevant to our theme have been discussed in preparing for and during UNCSTD", what is now needed is an "exercise in the analysis and assimilation of all this effort and a distillation of the essence of wisdom in it. In the face of the danger, however, that the valuable results of UNCSTD might simply be shelved, it is necessary to take the Conference's diagnostic conclusions and to formulate prescriptive orientations for implementing them." Any such orientations must meet four fundamental requirements, viz. those for complementarity among developing countries, internal consistency of proposed strategies, political realism and respect for each nation's cultural heritage.

Until quite recently, "the role of science and technology in bringing about significant changes in society has been considered... as a 'technical' problem that is to be dealt with mainly by the professional scientists and technologists.... [And] as a rather drastic oversimplification, we might say... that the social scientists are not well versed in scientific-technological practices, while the [natural] scientists and technologists are still rather insensitive to the socio-political implications of their activities, and even to the full extent of their economic consequences." There is thus an obvious need for an interdisciplinary effort aimed at formulating programmes for the realisation of desirable future transformations.

As pointed out by other speakers, such transformations must be seen within a global perspective. While the majority of the population of the Third World is living below subsistence level, it is quite mistaken to consider this state of underdevelopment as having been reached independently of events in the other 'Worlds'. The problems of developing countries, therefore, "cannot be discussed in isolation from the nature of current problems and developments in other parts of the world". However, "dialogues between 'North' and 'South', though important and necessary, cannot by themselves lead to a solution of the problems of the world". At a more fundamental level, "there is a need for persistent intellectual effort originating from within [each country] and leading to a specification of objectives and strategies as well as for the choice of alternatives". Coming close to the position adopted by Dr Despić on the previous day, Dr el-Kholy stated that "the crucial factor here is our ability to achieve socio-political systems that would enhance the efficiency of utilisation of their resources. Only effective forms of such systems [will be able to] provide the driving force needed to start and sustain the changes required to overcome underdevelopment on the national, regional and international levels. Intellectual effort plays a leading role in the realisation of such forms of socio-political organisation and is the only guarantee of the rationality of national and regional decisions." Such an effort should, in particular, "be based on the recognition of the specificity of the conditions in the Third World as a whole and within each country; [and it should place] emphasis on co-operation between Third World countries...."

"If, under the impact of the revolution in communications, our world has become very small indeed, this should not mean the obliteration of civilisations incapable of asserting themselves under the present circumstances. On the contrary, this should lead to their liberation and to the creation of a suitable climate in which they could provide humanity with the full richness of their heritage of thought, art and values. The solution we are seeking for world problems is a solution for all of humanity. Thus, it can only originate in the experience and heritage of all civilisations and countries. This is no call for chauvinism, nor does it mean that theories 'originating from reality' are the rejection of all that is positive in other civilisations and systems. Rather it is the realisation that neglecting other civilisations — past or contemporary — or failing to analyse them deeply so as to reveal the positive elements in them will only lead to more global problems and more underdevelopment and subordination. One of the more important positive elements in Western civilisation is the development of science and technology and the very close links that have been forged between them, while one of its most serious negative impacts is the obliteration of the civilisations of others", and especially the destruction of their cultural identities and life-styles.

Really, how relevant is Western 'culture' as such to Third World peoples today? "Through naïve acceptance of the superiority of the Western cultural model, we have tacitly adopted three basic assumptions: (1) that this development pattern is desirable in itself and is suitable for our society now and in the future; (2) that its realisation is possible to achieve nowadays as it has been possible to achieve in the past; (3) that our own experience so far in following this path is encouraging. The simple fact is that none of these assumptions is true, theoretically or empirically. Dissatisfaction with this model is now widespread within the industrialised societies themselves; the signs of its disruption and breakdown, materially and spiritually, are now recognised by those who

IV The Control of Space and Power

adopted it. This pattern was based on a reckless squandering of resources and disruption of the environment which is neither possible nor acceptable nowadays. [And] our experience so far is that adoption of this pattern has widened the gap between rich and poor, heightened social tensions and resulted in more dependence and subordination to the developed world, with grave political consequences that threaten world peace."

Yet, "the nature of scientific-technological activity and the role of science and technology are predetermined by the development pattern and life-style we choose. Adapting the Western model means that national effort will be restricted to the importation of technology from abroad, with its ready-made solutions developed by a far superior technological potential, for the satisfaction of a social demand for the goods and services that form the material basis of this lifestyle. The national scientific-technological effort will be geared to the needs of the élite, and it stands no chance of competing with the developed world in this race. At best, our scientists and technologists will be called upon to participate in some adaptive effort or, in the extreme, to imitate the production techniques that provide these goods. There is [in fact] only one viable option open to them, viz. to become integrated in the framework of a transnational corporation, at the latter's own terms. . . .

"Rather than allow contemporary science and technology as practised outside of our society to dictate our socio-political systems and to alienate us from our cultural roots, rather than let 'progress' be an outside force beyond our control, we seek an order within which alienation disappears, or — at least — decreases, and within which man becomes master of science and technology. . . directing them rationally towards the goals of harmony and equilibrium with resources and the environment, of satisfaction of essential needs, of justice and [the] liberation of Man's faculties on the basis of positive elements in our cultural heritage, and not [on the basis of] the dictates of profit maximalisation that currently prevail in international relations. This is the essence of self-reliance, reliance [based] on liberated creativity and sound traditions." Within such a framework the function of science and technology will change fundamentally: their role will be "to provide the technologies needed to bring about the alternative life-styles we may choose". Simultaneously, however, the possibility would emerge for an exchange of technology "as practised nowadays between the developed countries, rather than the unidirectional transfer from the centre to the periphery". In practical terms, this technological self-reliance can be characterised as the autonomous capacity to: (1) formulate policies, draft and implement national plans; (2) exercise well-informed social control over technology; (3) change and adapt imported technology; (4) exploit imported technology effectively in terms of socio-economic criteria; (5) innovate and deal effectively in the world technology market; and (6) maintain national cultural identity.

In proposing a course of self-reliance, "we are not seeking the return to a glory that has vanished. Such romantic ideas, usually tinged with sanctification of the past, make of our societies museums of culture and lead to extremist and reaction-ary concepts that ignore the weaknesses and defects that led to the passing away of those golden ages." "Development that is not the copy of another model or a slave to it is bound to be the conscious effort of an educated and well-informed society, enjoying freedom of thought and expression, unfettered by pseudo-religious obscurantism and intellectual bigotry." Yet it is essential to observe here that "such obscurantism and bigotry are usually veiled by the promotion throughout society of a view of science as a deterministic discovery of ultimate and immutable truths and not as an endeavour to understand better the world we live in. This 'magical' view of science, in a stagnant and autocratic society, leads to intellectual oppression and manipulation of public opinion. The label 'scientific' is used as the means for validating the views and values of the power groups in society. It becomes the justification for suppressing 'unscientific' opinions and the views of the 'laymen' and the 'extremists'. Let it be stated clearly here that what is at stake now is freedom for the whole of society to participate in the decision-making process and not simply a legal or formal definition of the rights of man, commendable and desirable as these may be."

Scientism also frequently takes the form of "an adulation and blind faith in the achievements of technology, which are presented simply as great victories of the human mind and of man's endeavour to master Nature. This masks the hidden forces that have motivated such developments, the physical and social problems [entailed] and the disruption they bring about; [it] presents technology as a disinterested and disembodied activity worthy of admiration. It renders acquisition of the products of modern technology synonymous with progress; and even in the field of armaments, technology is depicted in euphemistic language and with breathtaking glamour that hide the ugly face of death and destruction it brings with it.

"Integration of the social and physical sciences and technology thus becomes an urgent need. Technological activity must be viewed as essentially social action which involves the whole of society. The issue of scientific freedom becomes particularly crucial for the social sciences, since they often clash with the vested interests in society."

In the context of ever-increasing globalisation of scientific-technological activities, there are a number of phenomena which have developed according to a dialectic proper to the developed world, but which nevertheless exert especially strong influences on the developing countries. For example, at the ideological level, in spite of the "devastating onslaught by social scientists," there is "the desperate stand... [taken] by certain natural scientists who still maintain that theirs is a neutral and dispassionate pursuit of knowledge for its own sake and who disclaim any social responsibility for the consequences of their actions". On the other hand, at a much more material level, there is "the crippling distortion of scientific and technological activities – particularly in the developing countries – under the pressure of so-called 'defence' requirements", which have "siphoned hundreds of billions of dollars away from financing further effort where it is most needed and from 'transforming the world' by means of the

IV The Control of Space and Power

application of available scientific and technological know-how". Finally, in terms of material organisation, the emergence of 'big science' and the large-scale multi-disciplinary R & D establishment has also brought with it "a whole string of critical problems for the periphery". While 'big science' may be "cost-effective within the organisational framework of industrialised societies", it is "very expensive to establish and to run"; and, following in line with the general international division of labour, it has resulted in "extreme polarisation of scientific and technological activity and its concentration in the centre". This area of problems must be considered in close connection with the rise and operation of the transnational corporation as "the most efficient form of integrated techno-economic activity and as the main investor in and exploiter of technological innovation".

At present it seems that "practically all initiatives by the Third World to transform the prevailing world order have been thwarted and frustrated. One could almost go as far as saying that they have boomeranged and are most likely to become in the near future a means for entrenching dependence and subordination". For instance, efforts to modify the Paris Convention of 1883 (governing the patent system) and the call for an international code of conduct for technology transfer have been rallying points for the Group of 77 in their search for bringing an end to technological dependence. Likewise, construction of an infrastructure of heavy industry (e.g. metallurgical and petrochemical) and the development of engineering industries and national consultancy services have been accepted as recommended courses of action for building up indigenous technological capabilities. Yet there is now a growing concern that an internationally recognised code of conduct for technology transfer "would be of considerable help to the TNCs" and "would weaken the bargaining position of the developing countries". On the other hand, construction of indigenous technological capabilities along classical lines "might well lead to further qualitative intensification of technological dependence and subordination" with the metropoles "increasingly monopolising the decisive elements of R & D, engineering, finance, maintenance, etc., leaving the developing countries with control over the relatively low levels of the productive system".

In addition to these considerations, we are faced, in general, with "the rather bleak political prospects the eighties seem to bring with them. The dialogue between North and South is grinding to a halt and is now replaced by confrontation and open threats to use force for settling global problems. While the West seems unable or, owing to internal conflicts of interests, unwilling to grapple effectively with inflation and energy problems, the socialist camp is divided and at war. Détente has been degraded, and the price for Salt II seems to be an increase in armaments. Local wars have become daily occurrences in the Third World. Any serious attempt at transforming the world and any prescription for action must analyse these symptoms, take full account of them and look beyond them to the future long-term prospects of the global scene."

Despite this sombre outlook, it is up to the Third World countries to take the initiative and unite to solve their problems. How can this be done? "The last decade has witnessed an increasing interest in prospective studies under the influence of the threats of global problems..., [and] experience with these studies has emphasised the suitability of the regional approach." In a series of three appendixes attached to his paper, Dr el-Kholy drew on the experience of the Arab League institutions with such regional studies, in order to sketch an elaborated framework for effectively organising "a multidisciplinary and systematic approach for understanding and influencing [both] the processes of transformation" and the role which science and technology play in those processes.

The real practical significance of any such formal organisational approach, of course, "hinges on a theory of the sequence of development stages and of the international division of labour, which thus gives consistency to social and economic priorities". Theories of this type differ appreciably, but, on the whole, they tend to fall between two opposing views. "In very general terms, we may say that the first sees the problem as one of backwardness, primitiveness of economic structures and low returns on labour. This leads straight to the adoption of technological solutions concerned with the importation of modern technologies compatible with cultural development in the West and, hence, to adoption of its consumption patterns. It then concentrates on favourable contractual conditions or optimum adaptation procedures. The other sees the historical development of colonialism, subordination, monopolistic practices and economic penetration as the cause of the alienation of human labour from the technological environment. The economic structures that were originally dedicated to the satisfaction of the needs of the citizenry have been distorted to comply with production and consumption needs from without and to operate to the advantage of the stronger colonial power. This suggests the search for technological solutions that would end this alienation through a new economic structure capable in the first place of satisfying the basic needs of the whole population."

In the light of these two general social evaluations, it is possible to depict three prospective scenarios from which "a clear definition of the role and content of technological activity can be deduced, as society moves from one to the other:

(1) The *'consumer' society* is one which adopts a cultural and consumption pattern derived from the 'Western' model and in which technology is imported according to the criteria of commercial profitability for certain social groups. A primitive economic structure would still prevail as well as the phenomena of the 'extended' family and a weak local market. Some improvement in living standard could be achieved by means of exportation of raw materials and primary goods based on production processes involving a rather low level of division of labour.

(2) The *'productive' society* is one in which the cultural and consumption patterns are the same as before, but in which technology imports are based on appropriate choices, efficient operation and successful adaptation. The economic structure is now more varied and improvements in living standard come from a

IV The Control of Space and Power

higher level of division of labour. Dualism of the economy, rapid expansion of the local market and closer links with 'superior' industrialised societies are now common.

(3) The *'pioneering' society* is one characterised by an independent cultural and consumption pattern; in it, technology is the natural environment for human effort or the [body of] techniques necessary for a 'productive' society and an economic structure that meets the demands of the people."

In each of these three cases, as noted above, "the role of science and technology in transformation depends on objectives for the future, the definition of which has to be guided by a theory and a concept of development".

In concluding, Dr el-Kholy made a point that had been touched upon by many of the other speakers, viz. that concerning the necessity of linking the power of modern science and technology with the endogenous culture and creativity of the peoples of the Third World. As long as the social function of science is judged "as bigoted and oppressive" (i.e. in accordance with contemporary practice alone), it is obvious that the social climate will not be conducive to integrating science and technology into an attractive scenario for transformation. "However, hope for the future lies in a rebirth of the original attitude of the culture of the region towards science." Speaking as a member of the Islamic community, Dr el-Kholy said: "..the scientific method as we know it today is a product of our cultural heritage, closely intertwined with a religion that clearly recognised the universe around us as a source of knowledge and exhorted people to seek knowledge even in China! One might well think that the royal road to desirable transformation is an assimilation of our cultural heritage on a new − or is it an old? − level."

In his paper, entitled *Science and politics in a changing world*, Dr José Silva Michelena first analysed major socio-economic trends in the world today, and then considered certain problems of technological dependence and the possible solutions for such problems. Dr Silva Michelena began by pointing out that, in a long-term historical perspective, the world is now undergoing "the process of transformation from the capitalist mode of production to the socialist mode of production". Nevertheless, "it is a fact that most analyses of the present world crisis, be they Marxist or not, tend to concentrate [only] on what is happening in the capitalist world, thus leaving aside the unity, albeit a contradictory one, of world history today". Dr Silva Michelena, on the other hand, took what may be called a 'Great Power view of history': ". . . the basic dynamics of the present transformation of the world are determined by the dialectics between capitalist and socialist camps, which, without ignoring the internal contradictions in each one of them, are mainly determined by the specific objectives of the great powers of the capitalist and socialist blocs". What is the precise content of such dialectics? "Consideration of the objectives of the Great Powers, both capitalist and socialist, leads to the conclusion that the bourgeoisie, as the hegemonistic class of the capitalist system, has a primarily economic interest [according to which it] tries to prevent the expansion of the socialist camp; from this derives its need to combat [the latter] politically, militarily and ideologically. On the contrary, as socialist powers try to expand their influence to other countries, their primary objective is a political one."

The process of transformation, "of course, is not a unilinear one, nor can it even be said that the outcome is inevitable or predetermined. Options are open to the point that it is not possible to say what final form the socialist mode of production will adopt. The so-called socialist societies of today, from a long-term perspective, can only be regarded as incipient historical experiments from which a more definitive form will gradually emerge."

The present world crisis is by nature "a structural one". "The world division of labour, which began early in this century, but which accelerated after the crisis of the 1930s, reached its limits (i.e. the impossibility to increase profitability for private enterprises) by the end of the 1960s." Attempts to escape from this general crisis have been based on: "(1) technological breakthroughs, which provide both new levels of profitability and new opportunities for capital accumulation; and (2) increase in the proletarianisation of the world by means of organisational innovations which facilitate the exploitation of low-cost labour". In addition, since the early 1950s a "new thrust in the international economy" has been provided by the growing process of transnationalisation, "which eventually resulted in an expansion of the social division of labour to an international scale and a further concentration of the means of production in the hands of a few gigantic enterprises". Although transnationalisation has not been enough to counteract the "structural factors which provoked the world economic crisis", so far "only seven or eight underdeveloped countries are fully incorporated into the new transnationalist structures; it is therefore possible that the process of transnationalisation has only just begun".

"The main consequences for the underdeveloped countries of the process of transnationalisation are: (1) an increase of the role played in the economy by the State, which is not only performing the traditional function of the State, but is also assuming the function of producing material goods in leading sectors of the economy; (2) a reconcentration of income in the privileged strata of the population, ... determined [first] by the need to expand demand for products of the leading transnationalised sectors of the economy (usually such goods can only be purchased by the capitalist sectors of the economy) and [determined] secondly by the need to keep down the real salary of the workers in order to make [investments] more profitable for the transnational corporations, which could otherwise invest somewhere else; (3) the relative deterioration of the capacity to produce both industrial and agricultural products oriented to the satisfaction of the needs of the impoverished masses of the population."

Both of the last two consequences "inevitably lead to the discontent of the masses which sooner or later may explode in violent reactions, strikes or even... revolutionary movements". Meanwhile, the fact that the crisis is a global one "determines a deepening of class struggle in the developed countries", which, in turn, emboldens the labouring classes of the developing countries.

IV The Control of Space and Power

The Latin American experience reveals two fundamental strategies for controlling such social movements. The first consists of "establishing or reinforcing a social and political pact [uniting] labour organisations, governments, political parties, the military and the bourgeoisie". Such a policy can be seen in countries such as Venezuela, Mexico and Colombia, "where more or less democratic governments exist and where Social Democratic or Christian Democratic parties have great influence both in government and in labour organisations". On the other hand, the second strategy for controlling social movements consists of "establishing authoritarian regimes such as [those] in Chile, Argentina, Uruguay and Brazil".

When viewed within the general dynamics of world politics today, neither of these strategies is "intrinsically stable". As stated above, the world is at present undergoing not only an economic transformation, but "also a political transformation... which is perhaps as important as or even more important than the economic transformation". In line with his general views on the dynamics of the historical process, Dr Silva Michelena explained the nature of this political transformation as follows. "Since the mid-1950s (and the emergence of the nuclear stalemate or 'mutual superiority' between the United States and the USSR, the locus of confrontation between the great powers shifted from the equilibrium zones (especially Europe) to the periphery. From then on, any war of liberation or revolutionary war emerging in the underdeveloped countries of the world was likely to be transformed into an indirect confrontation between the above-mentioned Great Powers, provided that massive logistic support could be given by both of them. Since the US could do so around the world since 1945, the matter was reduced to the increasing capacity of the other Great Power to give logistic support to popular movements. Apparently, the Soviet Union today is able to give logistic support to revolutionary movements in Asia, the Middle East and Africa. These are the 'hot zones' of the world today. ...Simultaneously, multipolarisation of the world has increased both economically and politically, thus making it possible to establish new alliances and pacts in order to take better advantage of the new social division of labour on a world scale."

It is within the context of these general tendencies that one can explain recent events such as "the increasing inability of the United States to enforce the applicability of post-war pacts such as CENTO, SEATO, etc." and "the emergence of an organisation such as OPEC" ("... a phenomenon made possible by the strategic nature of oil, by the will of OPEC nations to back the organisation and, last but not least, by the increase in the profits of the transnational oil corporations"). Furthermore, apparently overlooking the example of alignments before and during World War II, Dr Silva Michelena cited "the rift between China and the Soviet Union" as the factor which "made possible the formation of cross-ideological alliances" such as those which emerged into view during the Indo-Pakistani war and the Angolan revolution. Moreover, the case of the "intervention" of Vietnam in Cambodia "revealed that conflicts between underdeveloped socialist countries can also happen. Along these lines, it is not surprising that closer links... are growing between the US and China. One cannot even rule out new and perhaps more formerly unthinkable ententes"; and "even a new entente between the Soviet Union and West Germany... may be quite possible within this rearrangement of the world".

At present "the most significant" element in world politics is "the expanding capability of the USSR to give massive logistic support. In the last 30 years, the Soviet Union has gone from supporting Korea to backing Angola and Eritrea with the help of Cubans. Whether it will continue to expand towards Latin America is yet to be seen; so far, the compromise reached during the missile crisis in Cuba still seems to be operative. However, one can hypothesise that it may not be so by the end of the century." In short, from these general trends Dr Silva Michelena drew the conclusions "that underdeveloped countries will continue to suffer political instability and that the probability of revolutionary successes in the 'hot zones' is increasing".

"It is obvious that, in the face of such trends, dominant capitalist countries need to foster new means of legitimising the present situation in order to reinforce the more orthodox ways of economic, military and political domination. We [would] like to advance the hypothesis that one such means is the creation of a new myth which could both revive and make more credible the idea that under developed countries can, in effect, develop within the capitalist system. It seems that science and technology are to play a key role within this new developmentalist ideology." As noted by Dr H. Vessuri at the ACAST international colloquium in Vienna, such a myth could usefully fulfil three functions: "(a) the process of qualitative intensification of technological dependence, which predominates in most of the developing countries, could be conveniently disguised; (b) neutralisation and progressive obstruction of the few attempts of underdeveloped countries to control technology imports and direct investments, such as ... the Andean pact regulations, could be hidden; (c) the strategies of 'global planned obsolescence' and technological domination developed by a few multinational corporations of some of the main OECD countries could be efficiently legitimised".

Likewise, "concepts such as 'appropriate technology', 'increasing capacity to negotiate', 'technology transfer', etc., which appear profusely in the jargon of developmentalist ideologists, are but good ways of obscuring the basic facts that: (1) the true obstacle to satisfying the basic needs of the masses lies in the present system of domination; (2) local bourgeoisies, allied with the transnational corporations, are using technology to increase control and domination of their populations rather than to better their standards of living; (3) the industrialised countries are, in fact, less inclined to share on an equal basis the fruits of scientific and technological development; [and] (4) experience shows that industrialised countries treat science and technology as commodities to be exchanged in the market of underdeveloped countries on an unequal basis".

What, then, can the underdeveloped countries do to change this situation? According to Dr Silva Michelena, as long as problems of science and technology

IV The Control of Space and Power

"continue to be negotiated only at inter-State forums, ... we can only expect millimetrical progress or no progress at all'' − if only because such discussions leave out "the most important factor... the transnational corporations". Therefore, "it seems to us that the time is ripe either to create a [new] specific organisation or [to designate] an existing organisation [such as the Group of 77 or the non-aligned movement] to adopt as a priority the objectives of dealing directly with the transnational corporations in a global way. Then and only then can a more substantive context be given to common efforts to increase capacities to negotiate, to create an information bank, to foster managerial capabilities, to create multinationals of the developed countries, etc. One reason why we think such an operation may work is that transnational corporations, as in the case of OPEC, may also derive benefits from it. Among other things, uncertainty could be reduced; and therefore they could plan future ventures and profits in a more reliable way."

Undoubtedly one of the most stimulating papers prepared for the conference was that by Dr Zoran Vidakocić, entitled *The technology of repression and repressive technology: the social bearers and the cultural consequences*. Unfortunately, Dr Vidaković was quite ill when the conference took place, and he was thus unable to address any of its working sessions.

Dr Vidaković began his paper by observing that one of the great cultural phenomena of our time is perception of the fact that "the social functions of science and technology have been *mystified*, refracted through the prism of the [dominant] ideologies and stated in the fetishised frameworks of productivism, economic 'growth', 'promotion of civilisation', 'technological solutions' to social contradictions", etc. This basic perception is the result of continuing socio-economic crises and of social struggles both in the Third World and in the industrialised societies. Dr Vidaković's paper is an attempt to elaborate on this perception in regard specifically to "the *technology of repression,* i.e. (1) armaments and their scientific, research and technological potentials, (2) the para-military sectors (nuclear energy, outer space research, etc.) and (3) the 'reserve potentials' of totalitarian control over man and society (biogenetic, psychological, meteorological, nutritional, etc.)".

By way of introduction, the basic points of this elaboration can be summarised as follows, in a series of seven theses. First of all, "together with militarised science and technology in the service of force and violence, science and technology geared towards greater exploitation of natural resources and towards economic, socio-political and cultural hegemony in international relations [all] comprise a *unified organically linked structure of repressive function*". Secondly, "the main machinery of exploitation and rule within individual societies in international relations is decisively moving towards combining the monopoly of the technology of repression in the narrow sense with other forms of scientific and technological monopoly geared towards repressive functions". At the socio-political level, the ruling classes are tending to regroup themselves in a hegemonistic nucleus which expresses and makes possible the combination of both forms of repressive technology and which tends to consist of the military hierarchy, the military-industrial technocracy, the managerial nucleus of the transnational corporations, and the corresponding political and banking oligarchies. Thirdly, the bearers of hegemony continue to guide scientific and technological development "towards the expanded reproduction of the total conditions and factors of such hegemony". They are thus exerting "an ever more intensive effect... on the social character of the productive forces of labour", while absorbing a predominant part of the total potential of the scientific-technological institutions and leaving "a decisive socio-economic, political and cultural mark on the majority of scientific work and the technological application of its results". Fourthly, socio-political restructuring "globally conditions decision-making", in such a way that "the effectiveness of repression becomes a top priority"; and "the scientific-technological complexes in industry, agriculture, communications, medicine, urbanism, etc.... [actually] *thwart* the investigation and realisation of alternatives urgently needed for the material progress of human life.... Grandiose, diabolical *scientific and technological selection* is carried out systematically at the expense of the primary needs and historically formed progressive values of humanity." Fifthly, a monopoly of political power alone is not enough to legitimate such a selection: also necessary is "a specific socio-cultural articulation", according to which the interests, values and motives of "organised knowledge" are established and structured as functions of the scientific and technological advancement of effective repression. Concomitantly, in order "to legitimise scientific and technological monopolies and their repressive aims", ideological and theoretical forms have been developed in the social sciences (e.g. "the functional systems theory of society..., neo-Malthusian crises theories, 'socio-biology' as defined by Wilson and Trivers, etc."). Sixthly, "in 'developing countries', inasmuch as the process of their emancipation has not prevailed, technology placed under the guidance of metropolitan monopolies takes on *multiple and potent repressive* functions; [it] becomes an *essential* and may become the *decisive* factor in conditioning their structural dependence. 'The transfer of technology' is transformed into the implantation of military-technical, techno-economic, socio-political and cultural instruments for extending and continuing dependency and underdevelopment. With the help of these instruments, a fundamental *technological inhibition* is established in the development of these countries: a fundamental and radical frustration of scientific research and technological advancements that would be oriented towards primary needs and development possibilities. The repressive scientific-technological monopoly is constructed in a dependent society as an armature of international and, consequently, internal relations of exploitation; [it is] an armature geared to overcoming strivings and efforts towards economic and political emancipation. As a lever for ensuring the continuance of dependency, [it] is built not only into the material-technical structure of production, but also into the *class-structure* of dependent societies. Local oligarchies and 'élites' regroup... in function of military-political and techno-economic transmission; [and they tend] to impregnate the 'cultural

IV The Control of Space and Power

assimilation' of the authorised parts of the dependent society with the mystification of the scientific monopoly and repressive technology." Seventhly, exposure and abolition of the repressive functions of science and technology is undisputably a *"common denominator"* linking efforts for "progressive transformation in otherwise differing societies and in regions with materially unequal and culturally specific possibilities, priorities and choices. This common denominator is the global, *international* premise for the emancipated and autonomous, progressive and creative, contribution of all parts of the world" not only to their own scientific-technological progress, but also to the realisation of such progress on a world scale.

Analytically based, socially motivated and culturally articulated criticism of the repressive functions of science and technology can serve as the basis for deriving a profound long-term strategy for progressive transformation and development, a strategy that will aim at giving "the totality of scientific and technological development a significantly different quality". But one should harbour no illusions that such a multifaceted transformation can be realised without protracted efforts, both creative and preventive; its realisation will undoubtedly require "an entire historical epic". Our strategy today, however, "must begin with criticism and removal of those negative characteristics of scientific and technological development in which are condensed the most extreme... defects of [those] antagonistic structures that are, at the same time, the *constitutive obstacles* to the investigation and realisation of socio-cultural alternatives". Although it might be argued that repressive characteristics are interwoven into every aspect of the fabric of modern science and technology, the approach here adopted will be justified if it can single out the essential bearers and expressions of repressive functions, thereby putting the struggle into a historical relation by focusing on that which it is both "possible and necessary to subject to criticism and change" in our times.

Within this focus, three key tendencies towards scientific-technological repression can be distinguished: (1) the growing tendency for scientific-technological development to be oriented primarily towards goals of military force and totalitarian control of society, and the inclusion of such development into the international system of accumulation and distribution of surplus value; (2) the growth of 'a scientific-technological monopoly' in the hands of metropolitan centres which dictates not only the direction of basic research, but also the application of results to industry, agriculture, etc. – specifically for the purpose of consolidating their "economic and social hegemony in international and interregional relations"; (3) the determination of scientific priorities and technological selection in production according to the aims of increasing "forced exploitation and repressive control of the behaviour of the labour-force". The mutual interaction of these three phenomena constitutes "the dynamic of the expansion of the repressive, exploitative and destructive effects of science and technology".

One can in turn enumerate five main aspects of this dynamic:

(1) "The military-repressive orientation of science is transformed into an

essential economic factor of hegemonistic expansion and exploitation in international relations", since "the linkage of military and scientific resources creates an exceptional economic advantage: the metropolitan centres that gain this advantage will continue to expand resources and to control an ever-greater part of world accumulation."

(2) "The economic mechanism of international hegemony and exploitation. . . is established and functions in the presence of the international machinery of non-economic compulsion that in part ensures the reproduction of the social conditions for monopolistic accumulation. . . ."

(3) In the face of social movements for the emancipation of subordinated countries, "a specific scientific-technological control apparatus becomes increasingly important" as a means for protecting the threatened international order of hegemonistic expansion and exploitation; and this apparatus likewise "leads to profound inversion and distortion of science and technology with regard to the primary needs and developmental possibilities of societies".

(4) There consequently ensues a division of the economy into sectors with essentially different priorities for reproduction. Those sectors that do not enjoy the advantages of militarisation see their structurally limited possibilities for research and innovation "directed primarily towards the development of means of labour, of the technical organisation of the labour process, and of technological approaches that enable intensive exploitation", especially of unskilled labour. "In this way, the orientation of technology as an instrument of increased exploitation of labour is strengthened", while "backward and repressive technology in the production of goods is the other side of a system of production and accumulation that favours rising technological and scientific development in the production of arms and other means of repression".

(5) Together with other factors, monopolistic technological control of the dependent countries determines a situation in which "the development and use of science and technology are carried out within the framework" and for the explicit purpose of "the super-exploitation of the labour force of the dependent countries".

Taken all together, these five aspects form a vicious circle in which "the economic function of the production of arms stimulates the scientific-technological revolution in the military-industrial sectors, reduces the accumulative capability and the possibility of essential technical innovations in other sectors and orients [the existing] technology [in these sectors] towards forced exploitation of the labour force".

One of the most important manifestations of this dynamic is to be found in its effects on the structure of the international division of labour. "If the possibility for . . . exploitation of labour in metropolitan societies is limited, sectors of production whose capacity for accumulation is in danger move to countries in which they can create the conditions for more intensive exploitation." This exportation of sectors is effected from the developed countries towards the developing countries; but it also occurs from the more powerful to less powerful industrialised

IV The Control of Space and Power

countries. Worsening conditions, in turn, lead to opposition and resistance both in the Third World and in the second-class industrialised nations; in both cases, "the bearers of international hegemony react by strengthening the repressive apparatus" (in ways 'appropriate' to the conditions of each country).

Frequently "mystified as a scientific and technological *'gap'* between developed and developing countries", the international scientific-technological *monopoly* is in reality "an essential part of the system of international monopolistic accumulation and control of the conditions of production, exchange, distribution and consumption"; and "the repressive functions of science and technology are *directly* based on the *monopoly* of scientific research, the monopolistic private ownership of scientific knowledge and the exclusive control of its technological applications".

But all this is only part of the story. "For more complete knowledge it is necessary to shed some light on the totality of class, socio-political and cultural phenomena" both in the hegemonistic centres and in dominated social environments. In this respect, "the international scientific-technological monopoly is formed and carries out its repressive action especially by means of two basic social figures which represent a condensation of the international totality of antagonistic social reproduction: the metropolitan monopolistic technocracy, and its subordinate [satellite] local 'oligarchies' and 'élites' in dependent societies". The monopolistic technocracy "synthesises interests, motives, values and goals, forms of social organisation and a hierarchy of functions and positions that inspire intellectual production ...; at the same time, [it] directs this production towards investigation and selection of the optimal possibilities for exploitation and repressive action". It likewise "interiorises the repressive functions of science and technology and these functions determine its 'maximum consciousness'". The other social component of the international scientific-technological monopoly consists (in the subordinated countries) of dependent groups in symbiotic relationships with the metropoles and characterised by a "socio-cultural lobotomy", i.e. by a "subordination of interests and motives, an incapacity for research and knowledge beyond the framework of the hegemonistic interests and the models that are the incarnation of these interests...., [by] a caricature-like imitation of metropolitan status and cultural patterns, a basic insensitivity to the interests of the working masses in their countries and a scorn for the native culture of these masses, for their creative and productive potentials...".

The metropolitan technocracy was called into being by the militarisation of the economy and of science; and one can even say that the corporations in the technologically leading branches of production (aeronautics, electronics, nuclear technology, industrial computers and information systems, chemical industry, etc.) served as the birthplace of this technocracy. "Galbraith's image of the concentric circles of the 'technostructure' [in fact] most closely corresponds to the leading corporations in the militarised economy." Such corporations are characterised by the high organic content of their capital, by a high concentration of scientific potential and educated technical personnel – and thus by "profits that devour a lion's share of the surplus value". In these corporations were found the greatest possibilities for the social integration of a relatively broad circle of participants under the hegemony of technocratically reorganised leading groups which mediate between financial centres, State military programmes and the resulting subsidised development of science and universities. In order to constantly exploit new scientific-technological resources and thus to permanently maintain military orders, these corporations "introduce the appropriate technology of power and management, bring their managerial groups into conformity with these demands, and... construct a new symbiosis of class interests and status between ownership and technocratic management". Furthermore, "the consequences of the symbiosis of the military repressive system and the authorised monopoly [e.g. of science and technology] are well known with respect to the personal union and rotation of leading managerial groups on both sides, with respect to their combined influence on fiscal and economic policy in the interests of a militarised economy and a global military-political strategy, [as well as] with respect to the entire political process in the metropolis".

Although the technology of violence and death actually becomes the main concern of the leading metropolitan corporations, these militarised corporations nevertheless extend their domination over the totality of social production. Power over death and power over life are concentrated in the same hands; they are made to serve an identical purpose, and they are judged by identical criteria. The only difference between the two is that the use of their power for destruction and death is "significantly more effective" than its use for furthering survival and the relief of those human problems that are, in fact, consequences of this whole system of production itself. Within this context, it can also be pointed out that the militarised apparatus of production was the original laboratory for perfecting the so-called "scientific-systematic" and technocratic forms of rationality; and from it also was inherited "the basic irresponsibility and superficial 'political neutrality' of researchers and technical operators".

Within subordinated societies, on the other hand, there are two main generators for the formation of 'technocratic élites'. The first is "the local repressive apparatus (primarily military. . .)". This apparatus maintains itself "under the influence and control of metropolitan systems, thus carrying out perhaps the most important transfer of [military] technology and contributing to the formation of a specifically authoritarian-technocratic ideology", which "postulates *total repression* as the sine qua non for the survival and development of [any given] 'backward' society. . .". The second generator of local 'technocratic élites' is "a complex of 'technical aid' projects" which are "carried out within the global strategy of the metropolitan monopolistic centres". The selection, education and indoctrination of the cadres for such projects is performed either by "metropolitan educational and research factories" (and their local branches in the Third World) under the auspices of "superficially independent foundations"

IV The Control of Space and Power

or by the transnational corporations themselves within their own personnel-training programmes.

These two types of local élites typically remain on their own separate lines of development until a crisis hits. Thereafter, unless such a crisis results in liberation from the metropolitan system, the two tend to merge into a single restructured class-fraction: "these two components — the military group and the social arm of the transnational corporations — produce the local force that becomes the interested, politically and culturally conforming recipient of (scientific-technological and) total hegemony. Between these two components there develops a symbiosis of power and interests, an osmosis of ideas, values and orientations. The ideology of total repression unites with the ideology of technological and cultural dependence and assimilation; and in this union, repression gains strength as the condition for all (dependent) economic growth, technological progress and 'modernisation of the society', as a circle of insurmountable dependency based on the importation of prefabricated knowledge (together with technical and consumerist models) is closed up by the ambitions of the protagonists of authoritarian rule."

In presenting his paper entitled *Nuclear energy in Latin America: the Brazilian case*, Dr Luiz Pinguelli Rosa first of all made clear his general ideas about what constitutes the most suitable sort of general energy policy for developing countries. Dr Pinguelli Rosa observed that "an effective energy policy cannot be limited [merely] to meeting demand, nor can it be guided solely by the search for a minimum price. [It] must also *orient* energy demand so as to make it consistent with the global objectives of the country." In the case of a country with an inadequately articulated industrial base, but with an energy sector controlled by the Government, a correct energy policy can be a key factor in reducing the importance of foreign-dominated sectors and of balancing the general structure of production. Since energy enterprises are responsible for a substantial share of all purchases of equipment and since they likewise influence industrial costs by means of pricing, a correct energy policy can be an important driving mechanism both in national and in regional industrialisation. Such a policy in a developing country must as much as possible shift energy consumption towards indigenous energy resources, in terms both of technology and of supply. By this means, foreign currency can be conserved and security of energy supply safeguarded.

On the other hand, "energy consumption is nowadays large enough and concentrated enough to produce serious effects on the ecological equilibrium of certain regions. This entails a social cost which can be large, but which has in most cases been overlooked." Any adequate policy, however, must take this social cost into account: "otherwise the country will sooner or later have to pay for it".

Even apart from its ecological implications, it is clear that "energy policy is closely linked to social and economic policy. There is no way to separate it from the national planning we have in mind for the future. Either we will maintain a high concentration of revenue, socially and regionally, or we will try to reach a more reasonable distribution of the national revenue. This is not a rhetorical or an idealistic point. [On the contrary,] it would be unrealistic to separate the technical and political discussion about energy from its economic and social context." Institutional changes in the direction of democratisation and decentralisation are necessary here as elsewhere, and working people must make their voices heard in discussions about energy policy. In this regard, it should be noted that at present the domestic consumption of energy for different social classes is highly unequal ("sophisticated goods incorporated typically into the middle-class standard of living have a high energy content"), while means of public transportation are insufficient and their services poor ("the private car has all the privileges"). Social discrimination is also to be witnessed in the consumption of energy within the industrial sector, "which produces goods for a relatively small part of the population or for export, while neglecting the needs of the majority of the people almost completely". For these reasons, "the reorientation of energy demand is *the basic condition* for an effective energy policy in Brazil", as well as in many other Third World countries.

It is within the context of these general considerations that specific energy policies of given developing countries can be evaluated and guidelines for the formation of policy drawn up.

As a practical example, Dr Pinguelli Rosa analysed the case of the Brazilian nuclear energy programme, for which a certain amount of background information should be kept in mind. For example, "Brazil has a hydroelectric potential of 200 GW, of which only 25 GW is at present utilised; and it is expected that 150 GW will be used by the year 2000. In spite of the great distances between many of the waterfalls and the big cities, it is possible to transmit electrical energy with final cost of a hydroelectric kW at less than half of the nuclear kW cost. Besides this, there is coal in the south of the country. Therefore, nuclear energy is not an economic necessity to Brazil yet."

Nevertheless, "the Brazilian nuclear programme foresees the construction of eight light-water PWR-KWU reactors of 1300 MW each" by 1990, in addition to the 627 MW Westinghouse reactor now in the final stage of construction. Furthermore, according to a treaty signed with West Germany in 1975, Brazil has undertaken the establishment of an industrial base for the production of heavy equipment for reactors and for enriching and reprocessing nuclear fuel. Why does a developing country implement such an 'ambitious' programme? There are, in fact, a number of reasons. Some of them are "related to the long-term security of the country's energy supply, after the exhaustion of the hydroelectric potential and of other sources"; but there are also other reasons "related to the myth of nuclear power as a magic key for the progress of the nation...".

According to Dr Pinguelli Rosa, "nuclear energy may be necessary to the economy of the less developed countries in the future"; and within thirty years it will probably have to play an important role in supplying Brazil's energy. "No matter how great the Brazilian hydroelectric potential may be, the day will

IV The Control of Space and Power

inevitably come when this consumption will exceed that potential. Thus, Brazil cannot ignore nuclear technology, because it may need it in the future." This becomes all the more clear when one considers that an economical utilisation of solar energy for generating electricity on a large scale is "improbable at medium term". Likewise, "the myth that the developing countries have to concentrate on intermediate technologies" would here imply "the absolute priority of renewable resources and of rustic ways of energy generation", whose development still requires much time and investment. Such a choice would, naturally, be quite dangerous, for, among other things, it would mean abandoning native petrochemical and nuclear resources to exploitation by the rich countries alone.

On the other hand, however, usage of nuclear energy is a subject that demands a careful assessment of prospective risks and benefits and "a political evaluation that transcends the technical aspects". For such an evaluation, "the [proper] instrument is democratic discussion; and for this political discussion to be well-founded, the participation of the technical-scientific community is essential". Moreover, the Three Mile Island accident has highlighted the risks of accidents on reactors and the question of nuclear safety, while the problem of "the storage of radioactive wastes... stands without a final solution". Such problems "tend to be worst in less developed countries, for three basic reasons: (1) the necessity of adopting nuclear standards and requirements from other countries and sometimes from more than one country...; (2) the weakness of the national licensing authorities, which not only have small budgets, but also do not have the necessary independence and authority to fulfil some of their intended functions; and (3) the lack of well-established public opinion groups which could force the government into giving more attention to safety-related matters".

It should be stressed that "nowadays the acquisition of sophisticated equipment from the developed countries may not [always] be the most appropriate way to assure control of nuclear technology for the future". In this regard, the acquisition by the Brazilian Government of the German nuclear technology deserves to be severely criticised. Not only was this technology purchased at a very high price at a time when the country was by no means in urgent need of it, but also the project was badly adapted to Brazilian conditions and internal resources; in particular, it ignored the Brazilian scientific community "almost completely". In addition, the treaty governing the establishment of the nuclear-energy industry "fatally requires the importation of the equipment" from Germany; and even if eventually such equipment is partially made in Brazil, it "will be made by foreign companies or in joint ventures with them". Thus, despite the fact that the technology has been well-known for decades, the industrial production of equipment for the generation of electricity will see its already grave dependence on multinational companies intensified; and the foreign debt will be aggravated.

On the other hand, the nuclear power industry within the country will be under national control only when Brazil possesses all elements of the nuclear fuel cycle: "from this point of view the country must have both enrichment and reprocessing plants". Here arises the snag, however, for the plutonium produced by such plants is the essential material for making a nuclear weapon; and, on the grounds of restricting the proliferation of nuclear weapons, the USA is "trying to avoid the sale of a reprocessing plant to Brazil". Brazil now finds herself threatened with the abrogation, under US pressure, of the final stages of the Brazilian-German treaty. If these stages are not carried out, she "may find herself in the position of having bought the reactors and afterwards of not having any guarantee of fuel supply", while "relying upon an imported fuel (namely, enriched uranium, which may become more critical than oil) for a substantial part of [her] electrical energy".

Raising once again the question of collective self-reliance and advancing the proposals of the 1978 Interciencia Symposium, Dr Pinguelli Rosa suggested that in the face of such circumstances "countries at a comparable industrial and economic stage (such as Brazil, Argentina, Venezuela and Mexico) [should] unite themselves to develop a nuclear programme, eventually including other countries as soon as the latter need this kind of energy. . . . Continental co-operation would bring the enormous advantage of eliminating the possibility of a senseless nuclear race, with military implications. However, this co-operation would permit a rational internationalisation, at the South American level, of certain processes proper to nuclear industry, particularly the fuel cycle." "The USA, on the other hand, has proposed [another form of] internationalisation of the nuclear fuel cycle, in order to avoid the proliferation of such technology and the dissemination of plutonium." This internationalisation "under the hegemony of the countries already possessing nuclear technology", fails to inspire confidence, because of the "historical tradition of political and economic domination implicit in technological and industrial dependence. However, internationalisation at a Latin American level could be feasible and would overcome the objections currently made by the North Americans in relation to the fuel cycle. It would also make joint efforts possible, giving an adequate position to nuclear enterprises and strengthening the Latin American bloc in negotiations with the proprietors of nuclear technology." In fact (and at least partially in response to the attempts at obstruction by the USA), the development of a Latin American consortium to develop nuclear power "appears to be in the making". The USA has already been "virtually excluded from some South American nuclear markets" and is "still not considered a reliable supplier of fuel enrichment services and reactors". On the positive side, the consortium aims at linking the uranium resources of Latin America, the technical experience and expertise of Spain and Argentina, and the technologies available from Europe and Canada.

The question of the proliferation of nuclear weapons is, of course, an extremely important one, but it is imperative to place it within a proper context. It is true that the non-proliferation treaty has not been signed by many countries (including Brazil), who allege that it will only legitimate an unacceptable distribution of power, "by restricting the control of the pacific uses" of nuclear energy "without imposing any obstacle on the growth of the nuclear weapons of the military

IV The Control of Space and Power

world powers." On the other hand, Brazil *has* signed the treaty of Tlatelolco, which "forbids the production or possession of nuclear weapons and forbids the storage in the territory of a signatory country of nuclear weapons" belonging to other countries. It should be understood that in scientific circles many domestic critics of the Brazilian nuclear programme have, in fact, supported the Government's position in regard to the non-proliferation treaty. Why? Because "the objective of the scientists is for the country to follow an energy policy which is suitable to its real means and which leads to greater autonomy". On the other hand, the purpose of much international pressure (such as that exerted by the Club of London) is "to limit the autonomy of the less developed countries"; international pressure groups for non-proliferation base their position on the hypothesis that if developing countries manage to master nuclear technology, their "political irresponsibility... will lead to a nuclear war". The underlying assumption here, of course, is that "the responsibility of the world's military nuclear powers is enough to guarantee that a nuclear war will not occur. The historical tradition of some of these world powers, responsible for the worst wars and devastation the world has suffered, provides examples which refute this assertion." The point here is "not to defend the nuclear militarisation of Latin America, but to put the international question into its real context: the question of the nuclear disarmament of the world powers. From our point of view, the correct position is to repudiate the military use of nuclear technology in all countries of the world, while making clear that the great threat to the security of mankind lies in the nuclear weapons arsenal of the military world powers."

Discussion

The discussion for the fourth session began with an intervention by *Vice-Rector Mushakoji*, who raised four questions in relation to the presentations which had been made. He asked first of all whether, despite the importance of nuclear development for Third World countries, the noxious and wasteful effects of 'big technology' could not be avoided and whether it would not be possible to develop energy-saving technologies which would be less conducive to centralisation and technocracy. Secondly, he again raised the subject of informational technologies and asked whether (contrary, for example, to nuclear technologies) they might perhaps be a force for decentralisation or "an element of divisiveness among the countries of the centre" which might serve to intensify competition among them. Thirdly, how will it be possible to overcome the myth of the innate technological superiority of the West? At the ideological level, there is perhaps the possibility of collaboration between intellectuals in the Third World and those in the industrialised countries who are aware of the fallacy of this technological myth. At the political and economic levels, on the other hand, Dr Mushakoji said that the experience of developing countries in their attempts to gain control over the entire nuclear fuel cycle demonstrated the importance of contradictions within the 'trilateral' region; and he asked whether the transnational corporations themselves were not caught in such contradictions which could be studied and utilised. (As an aside, Dr Mushakoji also mentioned that labour movements within the industrialised countries are characterised by "an ambivalence", since they can act as effective allies either of the Third World or of the multinationals; this ambivalence must also be taken into account in formulating the strategies of the Third World countries.) And, lastly, he asked whether "the division in the socialist camp may create various . . . alignments which may not be in the direction of strengthening the solidarity of the Third World".

Next, *Dr Furtado* focused on some of the most important transformations in the structure of global power; and he suggested that we are now undergoing "a crucial moment in the development of mankind, at which the focus of power is being displaced". Five of the most important elements of the global power structure today are: the control of markets, the control of financial capital, the control of the sources of non-renewable energy resources, the control of cheap labour and the control of technology. It is clear that every one of these elements or areas of control is the scene of an important shift. For example, the only really expanding markets in the last five years have been those of the Third World; and, despite various hypotheses about the possible future resilience of markets in the developed world, "the hard fact is that the markets. . . . now being disputed by the international corporations are those of the Third World". Likewise, "one of the most important centres of financial resources is now [also] on the periphery, viz. the OPEC countries". Although it might be argued that "such [financial] resources are mostly being used by the banks of the central countries, etc., this situation is already changing"; and the use of these resources will necessarily "have an impact on the world power structure". In considering non-renewable energy resources, "the fact is that the sources of such resources in the centre of the capitalist system have been tapped and partially exhausted", and it is only too clear that conflicts over such resources now mainly concern "the sources of such resources in the periphery of the system". Increasing demand for such resources in the periphery itself will probably bring about new forms of friction and of confrontation. The supply of cheap foreign labour for the Western countries is now also at a turning point of one sort or another. "The centre of the capitalist system has developed very rapidly indeed during the last thirty years on the basis of cheap imported labour", but the countries which have followed this course now face a choice: they must either continue on the same road, which will necessitate "transforming the old societies", or they must refrain from this, thus "increasing social tensions at home". It is sometimes argued that this dilemma can be avoided by intensifying the use of labour-saving technology, but it is also often forgotten that such a plan "requires an exceptional financial effort" which may not be easily mounted. Finally, in regard to technological resources, it must be kept in mind "that technology is not [just] a stock, but is something that is being created [anew] every day . . . . This is the real weak point

IV The Control of Space and Power

of the Third World countries. Technological dependence is a reflection of the fact that we cannot really cope with this challenge in a short period." However, "people with financial resources, people with markets, people with other resources, they can also have access to technology, they can buy technology in certain circumstances." And, if one is able "to transform one's stock of technology" and to import the means by which that stock can be renewed (as the Japanese have done), then the problem of technological dependence has been solved.

When Dr Furtado finished making these points, the chair recognised *Dr Štambuk*, who agreed with previous speakers on the importance of linking technological development with changes in economic and political power structures. In this regard, it is important to realise that at present "capitalism is... taking on a new form [and] trying to recognise itself in a way which will help [maintain] it as a main system of exploitation...". Many apologies for contemporary capitalism have already drawn attention to the importance of transnational corporations in economic and social life today; and it indeed "seems that capitalism is entering a new phase, which could be called a 'multinational phase' or a 'transnational phase'", in which regional state apparatuses are tending to lose their importance, while new centres of power are being developed to accommodate more effectively the new means of exploitation. If this is true, "then is it also true that capitalism is developing a new kind of technology which would be appropriate to that kind of development?" If so, how can developing countries deal with this new sort of technology? Would it not be possible to combat the influence of the multinational companies over new technologies by implementing the suggestion of Dr Lefebvre and creating new forms of these technologies which would be more "participatory, in the sense that the workers themselves would take part in the process of decision-making, not just as co-chairmen or partners, but as real subjects".

A 'world vision' proper to the transnational corporations has already been around for some time, detailing the best ways for them to go about "organising their power and their decision-making processes" and to "dislocate" and control national economies around the world. Dr Štambuk noted that, according to this ideology, existing state apparatuses are often viewed quite "negatively", i.e. as having "negative effects on the economic stability of the world". Of course, there is a certain element of sense in all of this, for, economically, the transnationals are "growing at a rate of 10 per cent, which is more than that of any country in the world...;" their production is "very well organised", and "they are basing their power and their development on [this] production itself". But their great fault lies in approaching contemporary changes "in a static way" and in "forgetting people". In countering the power of the transnationals, it is necessary to identify and rely on the social forces "who are, by their [objective] position, ready to fight against that kind of social development". For example, it is clear that "the national bourgeoisie, army elements and other élite groups in developing countries" often eagerly accommodate and are, in turn, supported by the transnational corporations. The immediate producers and even white-collar workers in developing countries, however, can provide the basis for quite effective resistance.

Dr Pandeya, in turn, took the floor and spoke about possible strategies for overcoming the present monstrous imbalances not only of political-economic power, but also of scientific and technological knowledge. Citing an old Indian maxim, Dr Pandeya said that it is not wise to meet an adversary only with his own weapons and that victory in any struggle especially requires a creative use of one's own resources. In practical terms of scientific-technological development, this maxim signals the importance of mobilising the strong points of one's own heritage and people. As noted by Dr Hassan already and by Mr Blue later on, the major ancient civilisations knew astonishingly long historical periods of scientific and technical creativity. Despite distortions by contemporary Eurocentric historiography, these civilisations "created the first science that mankind has ever known; they developed it and converted it into a social-cultural-civilisational resource". And their societies were "probably the most forward looking and the most vigorous" of their times. In the last several centuries, however, these societies have for a number of reasons experienced a rupture or at least a waning in their tradition of scientific-technical creativity.

The proper question that must then be asked at present is: "How does one make a recovery?... How does this giant effect a re-awakening after a long sleep?" And how can the fostering of science contribute to such a re-awakening? According to Dr Pandeya, if science is to become a force for this kind of transformation, then the scientific culture necessary will involve much more than simply textbooks and research programmes: "this part of the package is transferable with a very marginal effort on the part of the nation. All you have to do is send out your people, get them trained, give them the facilities as Japan did and as all nations of the Third World have been trying to do. This is not the crux of the matter.... If the national political will is there, it can be accomplished in less than three decades of time, as India's case demonstrates. In 1947 we had practically no resources, today we have about the third largest body of trained manpower in the scientific and technological sphere [in the world].... This is not what really holds back." In fact, such a line of development can be "the most dangerous kind..., if you simply treat it as a dissociated, hermetically sealed capsule": it can simply amount to making a nation into "better material for total co-option into the new capitalism". If science and technology is really to flourish in the Third World, it is necessary to leap over this trap.

The real solution and the goal to be sought by all developing countries lies in the creation of an effective and genuinely popular scientific culture, i.e. a generalised social capacity for thinking and acting according to "objectively rooted" insights. "But this involves a capacity which the experts... of our crystal palaces of glitter, prestige, complacency and satisfaction will not generate." It likewise involves a capacity which in the last forty years has increasingly fallen victim to what may be called "communications pollution"; for, rather

IV The Control of Space and Power

paradoxically, the development of communicational technologies has tended to bring about "a total obliteration" of the capacity for large-scale independent thought.

Perhaps the real key to constructing a broad scientific culture lies again precisely in touching and re-awakening the deepest aspirations of a people. "In our cultural and civilisational history, we remember Buddha; not because the insights that he produced were fundamentally discontinuous and new, but because no-one has equalled him in this skill of converting insights into a people's resource, converting them into a cultural and civilisational resource, which went on operating for the next 1800 years...." This is, likewise, what Gandhi did. And this is "the kind of conversion that we need on a large scale, on a systematic scale, on a continuous scale – not only for India, but also for an entire three-quarters of humanity".

The next intervention was made by *Dr Maraj*, who observed that political leaders in the Third World countries are typically faced by "increasing demands for social services, health, welfare, etc."; at the same time, these leaders have difficulties in finding resources with which to meet such demands. In order to solve problems and thus to retain political power, many tend to "do deals with the multinationals... [and] with the larger countries, which more often than not come disguised as friends". The transnationals and the larger countries, in turn, tend to play developing countries off against one another in order to obtain the greatest benefits possible for themselves. This was the experience of the Caribbean countries, for example, when they accepted the tourism industry; and there is a strong possibility that this experience is being repeated now in the field of non-renewable resources. In this regard, the suggestion that developing countries should unite and bargain collectively in order to safeguard their non-renewable resources seems very reasonable. "The strange thing, of course, is that these countries are parts of other pacts, other groups. It does not seem that they themselves constitute a single grouping at any point in time, and therefore their weakness in negotiating remains [while] a tendency for history to repeat itself is likely to be very much emphasised...."

Dr Maraj said that Third World intellectuals "have a responsibility ... to educate" their political leaders, "who are not often either exposed or attuned to science and technology". Politicians must be informed, for example, of the key considerations to be kept in mind in negotiating with transnational corporations. Populist leaders in the developing countries are often "predisposed to make gestures to the rural people" and to be satisfied with "low-level technology". Yet these leaders must be made to understand that their countries require "the most sophisticated technology" if they are to take their rightful places in the twenty-first century.

In the next intervention *Dr Rasheeduddin Kahn* first of all stressed the importance of having a clear understanding of the differentiated structure of the Third World. The Third World contains something like 70 per cent of all mankind and constitutes "a most differentiated mosaic of large and small countries, spanning three continents and Oceania, [situated] at various levels of techno-economic development, reflecting a wide variety of cultural backgrounds and histories, ... [and] organised in many forms of political cultures and systems...".

Dr Rasheeduddin Kahn also stressed that in the contemporary world it is not possible to speak realistically about scientific-technological transformation "without bringing in the most critical factors of the State and government as instruments mediating [the utilisation of] science and technology for change". Since World War II, States have tended to become "maximal States". While "the whole concept of laissez-faire States [minimal States] is obsolete" and "the ideologically charged pejorative term of 'totalitarian State' ... is highly polemical and misplaced", in fact, a large number of metropolitan States today function as "total States" with a thick finger in every pie, including that of science and technology. Given this fact, there are three options open to Third World governments seeking to develop science and technology; "the first option is to accept client status... to one or the other major industrial power.... The second is to depend on transnational corporations not only for the transfer of technology, but also for investment patterns and service structures. And the third is to work out a co-operative regional arrangement (for pooling resources, talents, etc.), together with mutually beneficial collaborative arrangements with certain countries on a bilateral or multilateral basis", as has been encouraged by all recent summit conferences of the non-aligned movement. "If a certain measure of national dignity and creative innovation... is to inform the process of change", there is no choice "but to opt for the third option". There is, in reality, no fourth choice — for "while the IMF and the World Bank appear as autonomous institutions of a multinational character supported by the United Nations system", they are actually to a large extent working "under the auspices of the transnational corporations". Dependence of the developing countries on these corporations or on foreign governments means sacrificing national economic welfare, allowing the spread of an inevitable net of corruption and, in turn, allowing a black market to flourish. In this respect, as an Indian MP, Dr Rasheeduddin Kahn mentioned that in 1974 the Indian Finance Minister had "candidly admitted on the floor of Parliament that the black market economy in India was almost equal to the total [official] GNP". Because of factors like these, it is impossible to separate the scientific-technological prospects of a country from the broader political orientations of its state apparatus.

In concluding his intervention, Dr Rasheeduddin Kahn posed to the conference several major questions which were on his mind. Firstly, "how do we integrate the potentialities of modern science and technology, originated obviously within Euro-American civilisation, with the given sociocultural situation in each specific country, region, state?" Secondly, "how do we circumvent the adverse political consequences and cultural impacts of importing investments, [partially obsolete] technology, ... and management patterns through the mediation either of the

IV The Control of Space and Power

transnational corporations or of the industrially more advanced countries...?" Thirdly, "how do we remain contemporaries in a highly interdependent world... which is most unequal in terms of techno-economic development?" In other words, which means must developing countries adopt in order to become up to date in terms of world scientific-technological achievements? Fourthly, after the long slumber of colonialism and after the many battles against suffering and exploitation, "how can we use our national political sovereignty... as an opportunity and a challenge to bring about the much-delayed completion of the unfinished social-economic revolution?" And lastly, how can those countries which missed the opportunity of participating in the first industrial revolution "avoid being left out of the second industrial revolution" now under way?

Dr Issa then addressed himself to Dr Furtado, who had previously taken the position that the Third World countries are in the process of gaining control over several of the world's major power resources. Dr Issa, on the contrary, said that until the dimension of political power is taken into account, such control over these resources would be "doomed to remain only theoretical or, more precisely, potential". For example, in regard to the case of oil, it must be observed that the price of oil was increased in real terms only twice: in 1973-1974, in the wake of change in global power relations brought on by the October war; and in 1979, after the Iranian revolution. The point here is that "theoretical control over these power resources can only be materialised within a certain political context and by a rational use of the most important power resource: political power".

Another example is that of the Arab funds (petro-dollars), mainly invested in bank deposits and treasury bills in Europe and the USA, and estimated at about 90 billion dollars. "Since these funds are in dollars, the Arab oil countries are doomed to defend the dollar and the existing monetary system — in spite of the continuous depreciation of the value of the dollar and in spite of the obvious weaknesses of the present international monetary system — thus perpetuating the infernal circle of dependence." The most important fact here is that for all practical purposes these funds are blocked. "The Arab oil countries cannot invest them in the productive sector within the developed countries. The only possibility is to invest them in the Arab world and other parts of the 'Third World', But this possibility presupposes a completely different political vision and a different political setting within the Arab world." Dr Issa therefore asked whether it was really valid to speak about Third World "control of power resources within the present socio-political structure of the Third World and [in the light of] the contradictory interests of its parts and components".

Dr Le Thành Khôi next took the floor, to make two comments on the question of power. He first of all declared that "the basis of power is no longer the control of economic resources, of finance, of markets, of labour, etc., but the control of information". In order to illustrate this, Dr Le distinguished three phases in the expansion of imperialism. The first was "based on military superiority", but accompanied by "an ideological conquest" which boasted the 'superiority' of Christianity and the 'white man'. In the second phase the "central powers" dominated colonies with military force "but also with control of economic resources" and the international division of labour. Since World War II, however, industrialisation is taking place in dependent countries; and this is not only because of the profits reaped by the transnational corporations, but also "because it is enough for imperialism to control the industry of knowledge and information. To control the information of a country is to control its investment possibilities and thus, to be able to orientate its political and economic decisions...", not to mention its cultural and educational patterns. Dr Le's second point was that "the Third World is not a homogenous bloc"; for it contains both countries following a socialist road and those following a capitalist one. From this observation he drew the conclusion that "political differences are an obstacle to fruitful exchanges of knowledge between [the various developing countries] as well as to a united front in the struggle for scientific and technological independence".

In the final intervention of the morning, *Dr Abdel-Malek* summed up the results of the session and praised the position papers for demonstrating that, in discussing the concrete prospects for science and technology in the developing countries, "there is no way to go beyond the parameters of power". It is inadmissible to speak of present or future scientific-technological prospects 'in themselves', while abstracting them from their very specific (political and geo-political) contexts. Such contexts always frame one's possibilities: "we might not like it, we might dislike it totally... but there is no way, for example, to tell countries like India and Brazil..., 'opt out of the game of influence and stay where you are', ... 'keep on gliding with cottage industries and papaya growing". That won't do. No serious mind in those two countries will accept that'...". The aspirations of developing countries for up-to-date levels of science and technology must be understood in the more general terms of their struggles for a change in the global balance of power. "I don't mean the inversion of the balance of power, ... [for] it cannot be inverted for a long time; but I think it can be made more meaningful and more wisely spread over centres of influence and power." "There are centres of influence who have no power today; and there are centres of power who are losing their influence, who are still losing their own power at the historical level. ... This has been seen both as an economic weapon and as a moral curse: but it is neither the first nor the second. It is a factor which, if used within a radical political matrix, can... challenge the very civilisational model of the West; but if it is played by compradores, as it is in most cases now, it can reinforce the hegemony of imperialism. [However,] it is neither economistic nor ethical; it is political." In this respect, the fundamental question which was repeatedly raised before the conference and which faces all of the developing countries, despite the many great differences which distinguish them one from another, is how to live through the present transformation of the world "and yet remain sovereign cultural civilisations and political contributors, the subjects of history". Here we must face the fact that we live today in a world of States; and we must be realistic about the imposing role the State apparatus plays today. One may not like it personally, but this again is not just an ethical problem: it is a political problem. And one's evaluation of the role of different states should be linked to the major question: 'who are to be the actors of history?'

V

From Intellectual Dependence to Creativity

As observed by Dr Pinguelli Rosa in one of the last interventions in this Wednesday evening session, there was to be witnessed throughout almost the entire conference a marked ambivalence towards contemporary science and technology. It can be argued that this ambivalence was not merely a subjective attitude of those present: it was rather a reflection of the objective roles which science and technology are and will be required to fulfil. Thus, the various participants frequently seemed to be saying to themselves and one another: "Let us not be too presumptuous in speaking of 'the role' of science and technology in the modern world, for their functions are, in fact, variable and often contradictory." This point was brought home vividly by Drs Pečujlić and Vidaković when, in their position paper, they evoked the image of the 'two faces' of science and technology today; and implications for the social sciences were raised by Dr Bonfil Batalla who recalled the complicity of sociology and anthropology in the oppression of subjugated peoples. As stated so often above, alternatives for future developments in the field of science and technology hinge on the question of which social forces exercise effective political and economic power, and the alleged appropriateness or inappropriateness of technologies, for instance, should not be allowed to obscure the crucial question of *who* is doing the appropriating. Is it being done by alien forces or by endogenous ones? by exploiters or producers?

The functions and the rate of development of science and technology will fluctuate across a spectrum of different social contexts, but this does not mean that science and technology are mere epiphenomena superfluous to 'basic' social relations, or that they lack an objectivity proper to themselves. As pointed out by Dr Damjanović, modern science and technology are organic components of contemporary culture, affecting every society vitally through their impact upon production; and without a certain critical proficiency in generating scientific knowledge and technological skill, any society will be faced with stagnation and decay. On the other hand, the objectivity proper to science and (arguably) technology can be said to distinguish them from other forms of culture and to root them firmly in an international dimension, as recalled again by Dr Damjanović when he described them as the 'collective intellect' of the human race as a whole. Historical light was thrown on the differential aspects of scientific and technological universality by Mr Blue when he considered Dr Joseph Needham's work on

V From Intellectual Dependence to Creativity

scientific achievements in China and the West and by Dr Nakaoka in his detailed account of the mastering of metallurgical techniques in Japan during the nineteenth century.

It is the objectivity of the scientific-technological enterprise which makes possible relations of dependence in these fields. Thus, one can say that India at the turn of the century was dependent on Britain for the training of biologists or for rolling-stock, whereas one (nowadays at least) would not really say she was dependent on Britain for Christianity: she was rather subjected to a foreign domination, which included a certain religious aspect. This general domination also undoubtedly included a distinct and powerful scientific-technological aspect; between the two forms of domination, however, lay the difference that a people need not be Christian in order to flourish, but it must have an adequate scientific-technological base. An inability or impotence to generate the knowledge or skills required by a society places the possibility of dependence on the agenda; and, if such dependence becomes operative and is allowed to continue, it tends to make the impotence for dealing with changing endogenous requirements more acute. Such impotence can result, however, either from foreign domination or from an isolation from global achievements in science and technology; and, in fact, the two tend to go together, since domination increases the monopoly of hegemonistic centres over key skills and forms of knowledge, while isolation from global developments makes a society all the more vulnerable to domination by foreign powers.

Unfortunately, because the various non-European civilisations have to one extent or another been subject to European domination in general as well as to a dependence on modern science and technology (the development of which has for a relatively long period of time been centred within the European cultural area), there has arisen a fairly widespread notion that proficiency in scientific and technological matters is a uniquely European trait. This point of view was apparently a quite common assumption of many participants in the UNU Symposium on Endogenous Creativity held in Kyoto in the autumn of 1978, although Japanese achievements could, perhaps, have been expected to have offered convincing refutations of the thesis. During this fifth session of the Belgrade conference, special attention was paid to refuting it in the presentations by Dr Nakaoka and Mr Blue (in relation to the natural sciences and technology), in that by Dr Bonfil Batalla (in relation to the social sciences) and during the discussion in the intervention by Dr Pandeya. It might be suggested here that thorough criticism of the notion of 'European science' is necessary, because this idea is at present being employed as a powerful ideological weapon justifying the continued domination and dependence of Third World countries by portraying such conditions as eternal and inevitable.

What is there that Third World countries can do in order to achieve a creative capacity to generate the scientific knowledge and technological resources which their peoples so urgently need? We have already seen that, first of all, there are certain definite political and social conditions which must be fulfilled; these taken together can be said to be equivalent to the effective exercise of genuine national sovereignty. On the other hand, all five of the position papers presented to this session emphasised that, within the domains of science and technology themselves, it is crucially necessary to rely on and strengthen endogenous capacities and facilities, while simultaneously drawing efficiently on global achievements in all fields. The interplay of universality and specificity is essential to the development as well as to the diffusion of modern science and technology; this interplay must be reinforced productively, rather than weakened by the negation of one of the poles. Dr Furtado's final warning against romantically idealising the universality of contemporary science was undoubtedly made with the multinationals' flagrant disregard for specific national needs and conditions in mind; and his comment is thus very important at a moment when the peoples of Africa, Asia and Latin America are, to use Dr Abdel-Malek's words, running into an onslaught of scientism from the West. But this should not cause one to lose sight of the equally important fact that monopolistic structures at the international level work powerfully against universality as well as against specificity in both the development and the utilisation of science and technology. Dr Nakaoka demonstrated the importance of the interaction of endogenous and exogenous factors for scientific-technological advance, especially by peoples who are relatively backward in these fields; he stressed the fact that advances should be conceived of as occurring typically in leaps, and he illustrated from Japanese experience the point that such leaps in fields of science and technology usually go together with others in the social and political spheres. In this regard, Drs Pečujlić and Vidaković noted that real progress in the natural sciences and in the development of new technologies will necessarily have to be part of a 'great inversion' of the patterns and priorities governing social life in general; and they emphasised that the leading role, both in this general leap and in the scientific-technological one, must needs fall on the shoulders of the organic intelligentsia of the labouring people of each nation in the world. Dr Bonfil Batalla's stimulating account of the recuperation of the endogenous social knowledge of the American Indian peoples provided the conference with a striking example not only of the function of such an organic intelligentsia, but also of the vibrant creative and regenerative potentials preserved by even the most oppressed peoples.

Let us now consider in some detail the arguments heard during this fifth session.

Developing the theme of *Science and technology as organic parts of contemporary culture*, Dr Zvonimir Damjanović based his paper on the proposition that both human freedom in general and national liberty in particular ultimately depend on economic independence. In turn, the development of science and technology is a necessary prerequisite for such economic independence; and for this reason much attention is given to discrepancies in the levels of research and education between the 'North' and the 'South'. "Scientific knowledge, technological ability and organisational competence represent the real prospective potentials of a nation; liberation from classical, blunt forms of colonial rule, from foreign cultural domination and even from the insidious pressures of neo-

V From Intellectual Dependence to Creativity

colonialism will bring lasting fruits and stability, only when it leads people to scientific-technological competence, i.e. to the power to create and [to] develop their economy." This is not always sufficiently appreciated by those whose primary concern is cultural emancipation: their "dominant aim is to preserve and revive national roots in culture and so to open up the prospective of human civilisation as a plurality of national cultures". Because, in global terms, science and technology have long been concentrated in the metropolitan centres, they are "too often treated as something local ('Western science', 'European thought', etc. . . .)." Thus they are considered "foreign and alien" in other localities.

The real relationship between science and technology, on the one hand, and production and social life, on the other, is highly complex. "Basic knowledge maps onto everyday life – including production – in very complicated and usually unpredictable ways." The classification of sciences does not and cannot fit the division of practical aims. Rather, "being the 'motive force' of production *in toto*, science . . . acts *as criticism* of practical activities". Therefore, it is impossible to develop science simply on the basis of "local, divided practical activities"; and "a good scientific contribution to praxis is, in fact, usually the negation of a technique, its removal or exchange". Encouraging only local industrial interests works to produce routine improvements and not scientific breakthroughs; and although the integration of industry proceeds on a different plane from that of the integration of scientific disciplines, nevertheless, "only the integrated broader interests of industry tend towards science". On the other hand, "it is impossible to plan scientific application in detail; one cannot know in advance which people (and combinations of people) are most likely to solve a new problem on a scientific basis. A broad population should [therefore] be cultivated in science in order to achieve the competence of a society to develop." Two decades of Yugoslav experience in attempting to integrate science with praxis have shown that (1) success is possible "only if one starts with concrete definite projects reflecting real needs in praxis"; and (2) successful projects are as a rule carried out by multidisciplinary teams. An appropriate ensemble of different types of scientists is usually necessary to perform useful research, and a broad and active scientific social base tends to be much more inventive than a small group of narrowly specialised experts. Thus, one can speak of "a critical structure and mass" of the research medium without which science will most probably be unable to meet social and industrial demands.

Just as with mathematics and the natural sciences, a socially broad competence in technology must be considered an essential part of modern culture. It is often forgotten that automation was a characteristic of the *first* industrial revolution; the main innovation in present-day industry, on the other hand, can be said to be "the introduction of 'adaptive' machinery". Thus, one can trace a technological evolution from the simple tool through the programmed (or automated) machine which controlled a process to the adaptive machine of today. Examples of this last type not only perform set functions, but also "govern the organisation of processes, not only in industry, but also in many . . . services". The prototype here is, of course, the computer and the computer network so reminiscent of neural networks. The object of technological research is now often "a complex, only partially defined system whose potentials, behaviour and characteristics must be investigated as if it were a 'natural', living creature". For this reason, technological research and development can no longer be treated simply "as a step *from* science *to* application" but must rather be conceived of "as a basic scientific activity". This becomes especially apparent when one recalls that technological innovation itself has made possible new areas of basic scientific research, e.g. in experimental mathematics. This new status of technology nowadays cannot fail to bring home to one "the stubborn fact that there will be no more easygoing [piecemeal] engineering: only a 'critical mass and structure' of modern technologists deeply rooted in theory can be a guarantee of complete cultural, industrial and scientific advance".

Yet such an advance necessarily has its social and intellectual implications. Modern science and technology are "a way of thinking" and of approaching reality: they are not "a set of recipes or of static information". Thus, "it is not possible for a society to benefit from science and technology without being exposed (spiritual tradition, creeds and national prejudices included) to their revolutionary influence on human behaviour"; in particular, the extension of modern science and technology *tend* to work against the various forms of élitism. "The scientific-industrial revolution both pushes many workers in all fields out of their old jobs and makes it possible to educate great numbers of youths to the highest levels. It not only makes society richer, but also education cheaper and technically easier."

Throughout history, science and technology have been rooted in the human race as a whole, and they are, to use Marx's phrase, the 'collective intellect' of mankind. "All attempts to ascribe them to any nation or group as a local achievement or characteristic can be proven false." "Reproduced, corrected, adapted to experience, in comparative autonomy (by being judge over itself)," science is "the most objectivised and critical activity of mankind. Of course, its real appearance can be deformed by certain carriers . . . , but its lasting, collective spirit transcends the individual and local . . . . In this respect it is important to note that the contributions of the scientists of Asia and Africa . . . have remained the basic part of science even during the ages when it was maximally 'Westernised', even when other aspects of the cultures of those nations were suppressed."

"Science and technology remain *common denominators of all* national cultures. Accordingly, every specific cultural complex must adapt to this fact." Scientific culture is able "to co-exist (but not to co-act) with different creeds and religions"; and "whatever the national culture, science and technology can fit in as complements". Yet, paradoxical as it may sound, "no national culture will survive unless it makes space within itself for the all-human complement of scientific-technological culture".

"Science and technology are prerequisites of emancipation and development,

V From Intellectual Dependence to Creativity

not only because of their impact on production, but also as decisive democratic factors; their proliferation will render the majority of people competent not only on technical, but also on social and political matters." Thus, "the old truth that man is something that has been *built* gradually turns into a new demand that he be rebuilt, re-educated, that he re-adapt perhaps more than once during his working age. He must act at that very level at which changes in the environment are generated – i.e. at the level of science and technology. And he will have plenty of time to offer to others, to society . . . [But] it is not possible to combine effectiveness in changing production with passivity and a non-democratic order. One can hardly imagine a man driven by scientific progress, keen and able to follow the changes around him, who would be willing to accept bureaucratic, dogmatic leadership. The integrity of the social system [then] . . . will be preserved only on condition of some type of genuine democracy and self-government." (By the way, one cannot fail to note in this regard that "the basic, ultimate problem brought about by scientific revolutions is not unemployment" – "which can be solved via corresponding social rearrangements".)

Although at the global level the gap between developed/rich and underdeveloped/poor countries is still growing, nevertheless "many facts suggest that it is objectively possible to bridge this gap". Almost all of the new technologies are becoming cheaper and cheaper, and the developing countries will soon be able to have control over any of them that they will want. In fact, "the only thing that is not growing cheaper is human competence"; and in the future "the real 'currency' will not be the mass of produce, nor even possession of the momentarily best technology, but the ability of people to move ahead, to change production and services. It is, therefore, equally possible for a developed community to drop down low – if they neglect scientific education and technological culture – and for an underdeveloped country to jump high – if they put highest priority on science and education." The second scientific-technological revolution now in progress is opening vast new horizons for human creativity; but "those who do not adopt modern knowledge and remain non-creative will have it worse than those who missed the steam engine". Hence, "science, technology and education should be given highest priority in national life, as well as a high priority and support in [the realm of] international co-operation". And as a precondition of success here, everyone should have the effective right to share in them and to contribute to them. According to Dr Damjanović, "objective hindrances to progress in this sense for some – maybe many – countries will be inferior to the subjective resistance of bureaucracies, minor groups with leadership ambitions, etc . . . Here again only democratic trends of participation by the majority in shaping the future promise to break the passivity, pessimism and resistances proper to . . . that part of our inheritance which belongs to the past".

Mr Gregory Blue presented his position paper entitled *Joseph Needham's contribution to the history of science and technology in China*, and he suggested that Dr Needham's work was of interest to the conference not only for providing a compendium of traditional sciences and technologies in one of the world's major cultural areas, but also for putting the historical development of science and technology into a global perspective. Thus, for example, "a correct historical perspective on [their] long-range development might . . . serve as a cogent critique of positions which still portray all scientific thought as an eternal and exclusive possession of Western civilisation". The first third of Mr Blue's paper listed in some detail the results of Dr Needham's research as published in his magnum opus, *Science and Civilisation in China*; and it is impossible to summarise those results here. Suffice it to say that they are voluminous and cover the entire pre-modern range of natural sciences and technologies, from mathematics, through engineering and chemistry, to medicine and pharmacology. Let us, then, pass on to the later sections of Mr Blue's paper and his oral presentation.

As noted by Prof. Hassan on the previous day, not only the Chinese but also the Indian and the Islamic cultural areas knew long periods of scientific-technological flourishing in the pre-modern era. In fact, "the level of science and technology, . . . in almost all fields, was much higher for the first fifteen centuries of our era in Asia than in the relatively backward areas of Northern and Western Europe". Consider only a few examples out of many: Chinese and Arabic algebra were world front-runners throughout the Middle Ages. Likewise, production of cast-iron was widespread in China for a millennium and a half before it was so in Europe; and for an equal period of time the Chinese were studying sunspots before Europeans even knew they existed. In the field of medicine, protein-hormones and steriod hormones were produced in the seventeenth century, while smallpox innoculation was practised on a wide scale since the tenth century.

Such advances did not remain confined within the borders of China; and many techniques, for example, were transmitted from China to the West. The three famous inventions which Francis Bacon considered to have most revolutionised the world all came originally from China. Gunpowder had great effects on the social structure of Europe in the late Middle Ages and Renaissance, for cannons could destroy the strongholds of the feudal lords; the magnetic compass had tremendous effects on navigation, world trade and the opening of world markets; while the introduction of paper and printing opened new intellectual horizons for European societies. Cast-iron came in the cluster of inventions that travelled across the Old World in the fourteenth century, again with revolutionising effects. Related to this was the transmission of the metallurgical blowing engines, one of the most important mechanical antecedents of the modern steam engine. Finally, the stern-post rudder was another borrowed invention which contributed to the growth of early capitalist society.

The transmission of techniques is easily demonstrated, but the situation is more complex in the realm of scientific theory. "When talking about the mediaeval sciences, it is often claimed that non-European sciences were not really scientific because they did not have theory"; they have for this reason been called simply 'an accumulation of techniques'. "This opinion is perhaps based on the fact that theory was much less diffusible across civilisations than concrete techniques were." Joseph Needham argues that the reason for this is that "medi-

V From Intellectual Dependence to Creativity

aeval theories were basically ethnic-bound. The basic categories of mediaeval sciences were formulated in narrow cultural terms understandable [in general] only to people in [one] particular culture and not transferable from one civilisation to another: e.g. in China there were basic categories such as Yin and Yang." Nevertheless, it is important to realise that "within mediaeval science (not only Chinese, but also Indian, Islamic and European) there were highly sophisticated bodies of theory: but it was mediaeval, not modern theory".

Many of the techniques which did not come from Asia into Northern and Western Europe were important "not only for helping to revolutionise production in mediaeval Europe . . . [but] also for bringing about the rise of modern science itself. Modern science took over many theories and many of the techniques from the traditional sciences, and many of its problems were formulated in terms of traditional ideas. It was by synthesising ideas from the various traditions – Islamic, Indian, Chinese, European, etc – that the Europeans in the Renaissance and thereafter were especially able to come to a synthesis which joined mathematics to an interest in nature and thus produced modern science. It is often admitted in conventional thinking that modern science owed very much to the traditional sciences of the ancient Greeks, e.g. to the Euclidean model of geometric proof, the Ptolemaic and Aristarchan analysis of planetary motion, etc. If we accept this, however, it is apparent that the qualitative leap which marked the birth of modern science involved much more than a 'renaissance' of indigenous 'European' elements. In the rise of modern science the Indian numerals, for example, represented a technical innovation without which the work of Galileo would be difficult to imagine. From India also came the conviction of the theoretical possibility of perpetual motion, transformed but evident in Newton's first law. Within mathematics itself, Descartes' synthesis of two formerly distinct disciplines – algebra and geometry – was certainly based on important contributions by mediaeval Islamic algebraists. In medicine a circulation-mindedness which was ultimately Chinese can be seen in William Harvey's holistic approach to anatomy and physiology . . . . Also from China came the crucial knowledge of magnetism, one of the most important problematics inspiring and informing the outlooks of many modern scientists, including Gilbert, Kepler and even Newton himself".

Why was China, for example, able to maintain a long scientific-technological vitality? And why did she subsequently not generate modern science? In considering such questions, the approach adopted by Needham has been "to investigate thoroughly the Chinese technico-scientific tradition, while consistently granting ultimate determinancy to social and economic conditions". Thus, "in Needham's view, the Chinese form of centralised, bureaucratic feudalism was much more conducive to innovation than either the loosely knit system of the Roman empire or the fragmented system of European baronial feudalism. For one thing, the position of the direct producers was not so precarious in China as in Europe; and consequently, on the level of ideology, disdain for manual labour (and, hence, for technical ability) was not so all-pervasive as under the Roman

Empire." Later on, of course, the constitution of modern science was linked with the rise of capitalism in Europe; and "it would seem to be the relative strength of China's bureaucratic feudalism which ultimately, and despite a long period of scientific superiority and technical innovation, hindered the emergence of modern science in China".

Having introduced the concepts of 'traditional' and 'modern' science, Mr Blue went on to consider their significance; he began by dismissing a frequent way of distinguishing the two. There have been various attempts to portray modern science as being defined by specific procedures such as controlled experimentation, empirical induction or systematic prediction. The history of science, however, shows that none of these is sufficient for defining modern science, because all three were practised within traditional Asian sciences. What, on the other hand, those sciences did not have was "first of all, an organic link between mathematics and natural knowledge; and secondly, a procedure for testing fundamental categories of thought. The various traditional sciences exhibited a typically regional quality in their basic categories; and they were also characterised by the fact that these basic categories were not testable. Modern science, on the other hand, frames its experimental results in terms of quantified hypotheses which are intended to validate or invalidate the theories employed at any given moment. In this sense, modern science is characterised, first of all, by quantification; and secondly, by its methodology of experimentation, which is aimed specifically at testing basic theories. The methodology of modern science, however, is not completely opposed to the methodology of the traditional sciences. It is, rather, a more productive approach to studying Nature, a more thorough approach to discovering and controlling natural processes. In this sense, the birth of modern science has been called 'the discovery of the process of discovery itself', i.e. the conscious and systematic appropriation of the method pursued, blindly yet steadily, by the various regional scientific traditions. But, in contrast to the regional traditions, modern science exhibits a tendency to become increasingly 'ecumenical' (or universal, if you will) in so far as: (1) its quantified methodology of controlled experimentation upon Nature can be understood, practised and developed by all peoples; and (2) the processes of Nature encountered by all people can be analysed and harnessed for human benefit by means of this methodology."

There is, however, a problem here, since the ecumenical potential of modern science was not at all fulfilled with the birth of modern science. According to Mr Blue, modern science has for a long time remained marked by particularly European experiences, European processes of production and European ideas. "This is the problem of the temporal disparity between the emergence of modern scientific methodology and the realisation of its ecumenical potential."

"At successive points in time different fields of study in Europe lost their traditional character as they were successfully subjected to modern methodology. For some time thereafter, however, the traditional sciences of non-European civilisations would continue (and in some cases still continue) to develop at their

V From Intellectual Dependence to Creativity

own pace and to preserve an important body of knowledge and technique not yet included in the focus of distinctively modern science. To the extent that such a situation prevailed, any given modern science could only successfully examine those problems which had already been incorporated into European experience and to such an extent modern science retained a regional character." If we consider the Chinese case, it can be seen that there was a moment (in Needham's terminology, the 'transcurrent point') at which any European science surpassed the level of a corresponding traditional Chinese science and that there was a necessarily later moment (the 'fusion point') at which all of the knowledge and ability to control Nature contained in the Chinese tradition was incorporated into a modern, ecumenical science. In the fields of mathematics, astronomy and physics, for example, the European knowledge surpassed Chinese traditional knowledge within the lifetime of Galileo, or about 1610. Because the two traditions, the European and the Chinese, were fairly similar, it did not take long for the Chinese achievement to be incorporated into the structure of modern science, of what was becoming ecumenical science. The fusion point was around 1640. In the case of botany, however, the transcurrent point occurred only at some time between 1700 and 1780; but the entire traditional Chinese material on botany was not successfully incorporated into the structure of modern science until around 1900. If, in turn, one considers the state of medicine and judges according to therapeutic success, then the transcurrent point can be placed a bit before 1900, but not much earlier; and at present there has not yet been a complete synthesis of the Chinese indigenous tradition and the Western modern tradition.

A caution should be sounded at this point in regard to what Needham calls the "ecumenisation" or universalisation of science; such terms are by no means meant to imply that science will cease at a particular moment in time. Science will, of course, continue; and ecumenisation simply designates the situation in which "regional barriers to scientific advance are overcome".

It can be noted that in regard to its traditional science and technology, "the Chinese case had several aspects which facilitated study. First of all, Chinese civilisation has fairly continuous historical records throwing light on scientific and technological activity. Also, China did not have to undergo any long period of direct colonialisation, and, hence, there was no long period of direct suppression of indigenous traditions." Countries which do not exhibit these rather favourable conditions will nevertheless be able to find a considerable body of traditional technical and perhaps scientific knowledge – even if literary sources are not available. "It is instructive here to look at the Chinese experience since liberation in 1949, in order to see the way the field of medicine, for example, was handled. As in other civilisations around the world, traditional medicine is a very rich field in China. With liberation and with the alignment of political structures to the needs of the people, it was soon found that the written tradition in China was only the tip of the iceberg": as a matter of fact, there was a great deal of traditional medical and pharmaceutical knowledge – scattered and dispersed among the people. "This knowledge had not previously been systematised; it had remained scattered and confined" within the context of traditional social structures and relations "as clan secrets, family secrets, guild secrets. Nevertheless, . . . by establishing a constructive three-way relationship between traditional practitioners, modern scientists and the masses of the people, in such a way that practitioners both of modern science and of traditional science would aim at serving the basic interests of the people, it was possible to systematise, develop and consciously integrate the traditional scientific knowledge which was still alive . . . .

"At the present time one of the important problems which faces countries in Asia, Africa and Latin America is that of the relationship to be established between modern science and the traditional sciences of these nations. If we formulate this problem in the terms described above, several scenarios might theoretically be distinguished for the short term in regard to those living regional traditions not yet incorporated into modern ecumenical science. The first is the situation in which the dissemination of modern science would be obstructed for any number of reasons while the regional science of the locality continued in its purely traditional form. The second is the situation in which a modern science (still to some extent Western-regional) is successfully disseminated while the regional scientific tradition is ignored, . . . or suppressed. The third situation is one in which conditions allow modern science to be successfully disseminated while the regional science is fostered, gathered, thoroughly collated and analysed by modern methodology. A variant of the third case is the situation in which both modern science and the regional science of a different locality are successfully introduced while the indigenous regional science is fostered. In the first case, many of the ecumenical advances already made by modern science would remain unavailable to the population in question, and there would be wastage of time and energy as the regional science attempted to deal with problems at its own pace alone. In the second case, modern scientific methodology would face problems without the benefit of useful insights available in the regional science and technology, and there would likewise be waste of time, energy and also money in dealing with problems for which such insights would be valid. The third case would be the optimal one of ecumenisation as well as for heightening the nation's ability to deal with the natural environment efficiently."

Dr Nakaoka Tetsuro's paper was entitled *Science and technology in the history of modern Japan – Imitation or endogenous creativity?*; and in it he showed concretely how imitation and endogenous creativity had gone hand in hand in the development of science and technology in Japan.

Dr Nakaoka noted that at the UNU conference in Kyoto (November 1978) there had been both an obvious discrepancy and an underlying similarity in the ways Japanese speakers and those from other Asian countries looked at Japan's industrialisation. The Japanese stressed the "shortcomings of Japan's rapid industrialisation, focusing on destructive effects of science and technology, such as water and air pollution, disturbance of the ecological system, destruction of

V From Intellectual Dependence to Creativity

traditional culture, etc .... Science and technology were generally considered exogenous, and were opposed to the value of endogenous traditional elements." The speakers from other Asian countries, on the other hand, saw science and technology as instruments for social change. "They looked for new models of social change which were not constrained by replication of Western experience, but dependent on the diffusion of science and technology among the masses for the eradication of poverty, ignorance, disease, superstition and the hegemony of oligarchic groups." They were "interested in how Japan had been able to assimilate exogenous Western science and technology so rapidly; how these forces had been utilised for Japanese industry; and how Japan had diffused modern science among the masses". Despite the differences between the two approaches, however, they both shared the view that science and technology had been exogenous to Japan, that they had been 'borrowed' and imported.

Are self-reliance and imitation really two opposite poles? Precisely this point must be questioned if we are to understand fully the relationship between endogenous creativity and imitation. It would seem that people who want to speed up their industrialisation have, rather, to reconcile the two. The question then is: "How [can a society] react to exogenous influences and ... develop potential endogenous abilities?" That the two do, in fact, go together has been shown repeatedly throughout history. As will be seen, the Japanese experience itself bears this out: Japan failed when trying simply to import knowledge, without taking into account Japanese conditions. And even Europe had borrowed and integrated, for "in the early part of this millennium Europe learned much from the highly advanced science and technology of the Arabic, Indian and Chinese cultural areas. This process included abundant examples of imitation and borrowing .... But, once rooted in European culture, these exogenous elements triggered off the energy latent in the European domestic conditions, and Europe began to develop rapidly."

Let us return to Japan's industrialisation and consider it more closely, starting in the middle of the nineteenth century. "On this subject we have an excellent pioneering work by Dr Tsurumi Kazuko ... Dr Tsurumi rejects the view which considers science and technology as entities independent of the culture of any particular society. Each culture has its own traditional ways of knowing and of making .... This means that there will be a conflict between all borrowed technology and the indigenous culture of the borrowing country, which cannot be overcome until the technology has become integrated into the culture." Dr Tsurumi investigated the conflicts between the indigenous iron-manufacturing technology and imported Western technology during the Meiji period in Japan. "This approach strongly recommends itself as a method of techno-sociology. Comparing the various conflicts brought about by the importation of technology into some countries, we can expect to find many keys to understanding the relationship between technology and social culture." However, in comparing China and Japan, Dr Tsurumi always seems to consider self-reliance favourably as a positive value and to refer to imitation in negative terms. Yet it would be impossible for developing countries to achieve industrialisation without any imitation or borrowing of technology.

Naturally, however, imitation is not enough, as the Japanese experience shows. In 1875 the Meiji government launched the first modern iron manufacturing factory in Kamaishi, under the supervision of a British engineer. For twenty years small furnaces had operated there, also built according to a foreign design, but without foreign engineers; these had fallen into financial difficulties, but technically they had been successful. The government nevertheless ignored this traditional technology and turned completely to British methods. The results were disastrous. After a hundred days they ran out of charcoal. After a while production was resumed using coke, but this resulted in congelation of the iron and coke in the furnaces, and so the entire plant had to be closed down.

Technological and historical research points to the following three causes of the failure. There was a large gap between the modernity of the technology upon which the new furnace was based and the old-fashioned way of producing charcoal; the location of the furnaces and the total transportation system were not well suited for supplying raw materials rapidly, and the design of the furnace itself was fundamentally defective. Furthermore, the operation was run by foreigners who failed to take the characteristics of Japanese domestic iron ore and coal into consideration. So a fourth and decisive cause for the failure should also be added, viz. the Government's worship of the West. This initial failure to establish a modern iron industry in Japan demonstrates clearly the dangers of importing technology without paying attention to indigenous conditions, and it likewise shows the advantage of domestic technology – namely, its prior integration with indigenous conditions.

If we wish to examine earlier attempts at creating a modern iron-manufacturing sector we can also turn to the history of cannon-casting. Here we encounter what Prof. Nakaoka calls the 'self-reliance/imitation model', which might prove to be a valuable example for today's developing countries. The reverbatory furnaces at Saga, Kagoshima, Mito, Nirayama, Tottori and Hagi, were all based on *one book* in the Dutch language. There was a drawn-out process of trial and error: only half the iron would melt, barrels burst at the first shot, etc. . . . . "But it must not be overlooked that, in the midst of innumerable failures, they made steady progress." Indeed, in only a few years all the initial problems had been overcome; and "by the end of the Edo period [1600-1868] they had made about two hundred [cannon], including three cannon with rifled barrels, which were the latest development in contemporary Europe".

"In spite of the innumerable failures, the speed with which they assimilated new exogenous technology seems to us astonishing." There has been much debate about the reasons for this speed, but the position adopted by Prof. Ohashi Shuji is of special interest here. "Drawing on his detailed studies of the late Edo iron metallurgy, Prof. Ohashi has shown three distinct stages in the formative process of Saga cannon-casting technology; each of these stages had its own counterpart in the European development." The first stage was the casting of

V From Intellectual Dependence to Creativity

bronze cannon. In Japan this stage lasted from 1842 to 1859, whereas European cannon-casting technology had remained at the bronze stage until the middle of the eighteenth century. In both places it constituted the historical basis for later iron cannon-casting. In Japan this second stage of casting iron cannon took place between 1851 and 1859, and it corresponded to European developments which took place from the middle of the eighteenth century to the 1850s. The third stage, dating from 1863, centred on the ability to make rifled cast-steel barrels; this stage corresponded to European development since the 1840s. "It must be noticed that, although each stage covered only a brief period of time, Saga had passed through exactly the same stages, and in the same order, as Europe."

In this development they relied not only on their own experience with bronze cannon-casting, but also on many other achievements of indigenous science and technology, such as the making of firebricks, the utilisation of water power, Japanese indigenous arithmetic and, above all, the domestic iron-manufacturing technology. Craftsmen had long been making arms, such as swords and guns, and agricultural implements, such as hoes and sickles, from pig iron and steel; and the temperatures of their furnaces were comparable with those of blast furnaces. Thus, 'the craftsmen had a significantly high level in the art of casting and forging; and they were well informed about the behaviour at high temperatures of melted iron and various other materials".

"Without solid support from indigenous technology and from their own experiences in the preceding stages, any attempt at imitation could not be expected to succeed." This much is beyond any doubt. But could they not have attained the same results without any imitation at all? "Surely, but perhaps only very slowly." Trying to imitate a Western model did, indeed, spur them on. "Exactly because their attempts to cast cannon were an imitation of exogenous technology, these attempts were accompanied by new and previously unknown problems. Solving these required a higher level of technological skill than the engineers had actually attained." Fortunately, the gaps they encountered were each time sufficiently small to overcome; but, because of the presence of these gaps, the increase in their abilities is best described as a series of 'leaps', rather than as a simple 'progress'.

Japanese technological development has known many such leaps, one of which is generally considered to mark the birth date of the modern Japanese iron industry: 1 December 1857 witnessed the first fire lit in the blast-furnace at Kamaishi, a charcoal-fired blast-furnace, again based on a design found in the single book mentioned above. Apart from the leaps, there were, of course, failures; but these, too, had their importance, for they prepared the Japanese engineers for their next leap. "This characteristic [i.e. a series of small leaps] in Japanese technological development is, I suggest, extremely important for the developing countries now. In so far as the developing countries aim to reach the same technological level as the developed countries . . . in a shorter period of time, their development plans must necessarily be designed as a series of leaps."

Social problems related to technological leaps should also be of interest for countries embarking on their own development. Technical leaps have to be seen in their full social and historical contexts. For, although in itself a technological endeavour, each leap was always an inseparable part of some historical social movement. "The first leap grew out of the agitation which began with the social shock brought about by the Opium Wars and the appearance of Western warships and which ended with the fall of the Edo Government. Many cannon cast during this time . . . were fired against the Tokugawa army as well as against Western squadrons. The second leap was, of course, associated with the great social change after the Meiji restoration; and the third with the international tension between the Sino-Japanese War and the Russo-Japanese War." Later, too, historical events remained the incentive for leaps. "Broadly speaking, it is true that Japan always succeeded in harnessing the nationalistic passion aroused by periods of agitation and in using it as a driving force for a technological leap. This is still true now. For instance, the Japanese leaders made full use of the oil crisis of 1973 in order to build up a feeling of emergency which they were able to turn towards the development of energy-economising technology."

In regard to nationalistic feelings helping to create a technological leap, an especially interesting period in the history of modern Japanese science and technology is that between the two world wars. World War I strongly impressed the Japanese with the virtues of science: more concretely, they had suffered different sorts of shortages because certain imports had had to be stopped, and they admired the Germans for having invented substitute materials in similar circumstances by means of scientific ingenuity. As the late Prof. Hiroshige pointed out, "the trend which began with this war was one of 'Science for Resources', which meant science for the ensuring of resources and for the invention of substitutes as well as the science of resource materials." The problem Japan had faced during the war was a sort of 'technological dependence' like that which can now be seen in the Third World. Afterwards, therefore, stress was laid on independence from Western technology.

"It is ironical to see that the dynamic interaction of exogenous and endogenous forces, which had operated quite well when Japan concentrated mainly on imitation of the West, began to operate rather destructively as soon as Japan aimed for independence from the West, perhaps because of too much weight laid on the endogenous side." "Calls for self-reliance and for Japan's own science and technology evoked too much chauvinism." Moreover, the social pressures arising from rapid industrial development were giving rise both to a radical-revolutionary labour movement and to an extreme right-wing nationalistic movement, and the government hoped to use the extreme right to suppress the left. "In doing so, it encouraged the chauvinistic, spiritualistic and Orientalistic elements of the right wing. All these elements were aggregated into the movement of so-called 'Japanese fascism', which was very effective in mobilising the masses for the war but extremely inhibitory for the development of science and technology."

"Subsidising fundamental research, promoting the development of indigenous techniques, improving the system of scientific education and diffusing scientific

V From Intellectual Dependence to Creativity

knowledge among the masses – these steady basic efforts resulted in raising the level of popular scientific culture and prepared the way for later development, but the vast popular energy was oppressed and inhibited by the mystical and chauvinistic elements of Japanese fascism. Only when Japan's defeat in World War II had brought about the democratisation of Japan and eradicated these inhibiting factors could Japan begin to develop rapidly. This experience may also be instructive for the Third World."

Another point which might prove to be a valuable historical lesson is that national effort in making a technological leap necessarily has repercussions on social structures. "Japanese historians of technology and science have talked a great deal about the 'biased structure' of Japanese technology. This concept has various implications, but, roughly speaking, it refers to a social structure in which only a few areas of industry have been developed to a high technical level, while others remain at a very backward, or even pre-industrial level. . . ." Japanese scholars usually explain this structure as a reflection of the structures of Japan's "armament-based capitalist industrialisation", but Dr Nakaoka considers it "a socially transformed mirror image of the technological leap". The point is that it seems "impossible for any developing country to make technological leaps at the same time in every industrial area". In itself, the gap between the advanced and the backward areas of the economy can play an important role in development. "If this gap is not too large, it may work as a stimulant for the more backward areas . . . . If, on the other hand, it is too large, it leads to conflicts."

A practical example of the potential negative effects of such a gap is provided by what are called 'the company-castle towns'. A modern company is set up in a traditional agricultural area, eventually dominating the area and providing nearly all the work there, much in the same way that a feudal manor did. In the same tradition, a feeling of company loyalty is created, since the company takes over the role of the native community. Through its power the company can also acquire strong ties with the prefectural government, not to mention the city mayor.

Citing a well-known case, Dr Nakaoka said that "many Japanese argue that there was a strong connection between the way in which Minamata disease was diffused and this regional structure". Minamata was one such 'company-castle town'. When the disease caused by the organic mercury from the company's effluent started to spread, the company denied the fact, tried to refute all scientific evidence of it and refused to take any countermeasures. "In this, the company received the tacit support of the Kumamoto prefectural authorities." Thus, the struggle against the disease became the struggle of scientists against the 'company-castle town' structure itself.

In conclusion to his outline of this aspect of Japanese experience, Dr Nakaoka said that he regards the technological leap "as an element of dynamic progress in society. In favourable conditions it can work as an excellent incitement to endogenous creativity; in other conditions it can become the starting point for serious conflicts."

In his paper entitled *La apropriación y la recuperación de las ciencias sociales en el contexto de los proyectos culturales endógenos,* Dr Guillermo Bonfil Batalla considered the relationship between the Western social sciences and the traditional knowledge of society possessed by non-Western ethnic groups, and in doing so he gave special attention to the Indian peoples of Latin America.

All developmental projects that are set up nowadays, whether conventional or alternative, require social investigation. Very often, political considerations determine which type of social investigation is carried out and which schools of social thought the investigators represent. "The political ideologies implicit in or ascribed to each current of thought identifiable in the social sciences are frequently assigned a determinate role of the final selection; this fact, together with the preponderance of a particular tendency in the institutions and apparatuses which sponsor and maintain such projects, works slowly to enforce the hegemony of a certain 'way of conceiving of' the social sciences so as to entrench a corresponding way of doing them." In this 'acceptable' way of working, however, "those who have no voice in the process whatsoever are precisely those who will be the object of study". Whenever it has been recognised that the people themselves might have something to say in relation to social sciences, they have only been thought fit to give 'data' or at most to act as valuable informants. "But it never seems to enter anyone's head that they might be able to contribute anything to the way in which social science is conceived."

Yet there is another approach to social problems which is founded on "the affirmation that all groups involved in any process must participate actively in it. This participation should be not only conscious, but also deeply motivated". Such creative participation can "bring into play all individual and social capacities, both in the conception and in the execution of activities directed towards development".

It is easier, however, to recognise the need for endogenous creative participation than it is to define its status within the social sciences. Can we validly speak of a 'social science' in every group that possesses a distinctive culture? This problem is "quite slippery," but it is not essentially different from that which we face when discussing agricultural technologies or medical practices. Except that the necessity of recovering social knowledge is much more pressing and indispensable than that of reclaiming many other areas of endogenous knowledge. "Socio-cultural alternatives for development cannot be conceived without implying the recognition and legitimacy of a distinctive model of society, and such a model can only be formulated by being based on a systematic and organised conception of what the society is, of how and why it is being transformed, of what its history has been and of what its options for building the future are. What is required, then, is a sociology, or, if you like, an ethno-sociology."

It is obvious that all societies have some knowledge of themselves and of the societies they have contacts with. This knowledge allows the members of a group to regulate and justify their conduct towards each other and towards persons not belonging to the group itself. In fact, every sort of social knowledge involves a system of classifying people and behavioural patterns in meaningful

V From Intellectual Dependence to Creativity 137

social categories. "For example, the kinship system has to be known by every member of the group in many Latin American Indian societies, . . . because it is a code of social behaviour that implies reciprocal and asymmetrical rights and obligations. The same can be said of the systems of power, of the institutions of collective labour, etc. . . ." There are also categories which refer to relations outside the group itself. "Within Latin America, Indian groups often refer to their own group as 'the human beings' or 'the real people'; outsiders, on the other hand, are classified in various categories which sometimes reveal an ethnocentric ideology." Such a body of social knowledge is not abstract, but is the result of concrete history; and many Indian peoples attach great importance to a knowledge of their history.

Besides being systematic, the Indians' historical and social knowledge is also dynamic, since it constantly assimilates new realities. But is this knowledge at all institutionalised? "In approaching this question, it must be realised that colonial domination wiped out the institutions and specialists which undoubtedly existed in at least the more advanced pre-colonial societies." As a consequence, "social knowledge is today maintained in a diffuse form, without explicit structuring". There is now, however, an upsurge of Indian political organisations which are struggling for political change and for recovery of the Indian identity with its traditional knowledge. These organisations seek to create a new society with a modern intelligentsia which will be able to update and modernise the traditional knowledge.

It may have been remarked that so far in this discussion of traditional social 'knowledge', the phrase 'social science' has been avoided. "In the Western tradition, the concept of science has a restrictive connotation that refers only to a particular type of knowledge which meets certain requirements: it aspires, for example, to be universally valid; it is an institutionalised knowledge; and it presupposes as a specific condition that science reflect not only on its object but also on itself." Within the social sciences, of course, the criterion of universal validity is quite problematical. Very often "ideologies approximate scientific knowledge quite closely"; and the two are perhaps never rigorously distinguishable. But be that as it may, the fact is that everyone has an ideology, and this alone does not make his research unscientific. It should rather be said that "the sole manner of avoiding ideologising subjectively is precisely to make one's own ideology explicit, so that the research itself as well as its results might be constantly understood in relation to it".

It is often said that the societies we are considering are prescientific. If this were true, it would be quite impossible to recover and develop their endogenous social sciences, and the only possible strategy would be to accelerate the transmission of the Western scientific tradition to those societies which had not generated it on their own. "The only difference would lie in the use of such knowledge, in its employment as an arm for liberation rather than as a constantly more sophisticated tool for ensuring domination."

We have to realise, however, that 'even' the Western-style social sciences have not attained the rigour and universality characteristic of the physical-mathematical sciences. It is not clear whether this is a consequence of the nature of their object or whether it simply reflects the current state of their development. Yet, in the light of this obvious lack of universality, "what kind of scientific knowledge is to be transferred to societies that do not have an institutionalised social science?" This question has very important implications, for 'social science' in Latin America has often served as the foundation upon which governments based their policies towards indigenous peoples, a policy aimed at de-Indianisation and at integration of the Indian peoples into the dominant society. "Is this the social science the Indian peoples have to make their own?"

It is true that colonised peoples have a strongly distorted view of themselves. Such a view has been imposed upon them by a colonial order which never stopped at physical force alone, and it has subsequently been strongly internalised. On the other hand, "the social sciences developed by the colonisers and placed at the service of maintaining the colonial order ignore or mask objective realities, and this is their ultimate function; but they also give a systematic account of significant characteristics of the subjected societies, while at the same time saying a lot about the organisation and functioning of the colonialist society itself". In fact, a good insight into the latter remains a major deficiency of the colonialised people. they need such insights in order to direct their liberation struggles effectively. "Ever since national independence, the great mass of the Indian population has been kept outside of modern sectors of society, subjected to archaic and brutal forms of economic exploitation and discrimination." An Indian could make 'social progress' only by abandoning his Indian identity and by integrating into non-Indian society; even then, he usually remained at the bottom of the social scale.

The educational system, too, has aimed at de-Indianisation; there are both obvious and subtle pressures. First of all, the lack of jobs in the Indian areas discourages graduates with secondary or university training from going back to their communities. But, on the other hand, the process of trying to assimilate into the dominant society is painful: it involves "hiding one's origin, abandoning one's loyalties, changing one's name, openly adopting new customs, etc... Despite all this, complete acceptance within the dominant society is almost never achieved". In the case of Mexico, "de-Indianisation via the educational establishments was begun in a systematic and institutionalised form several decades ago". As a consequence, "the training of Indian primary school teachers grew considerably: by 1970 there were about 4000; today there are more than 25 000, a thousand of whom have had university training." The political motive for training these people has been to use them as agents of de-Indianisation.

But recent years have given rise to an unexpected phenomenon. "Various trade unions and political organisations have arisen which are made up of and directed by young Indians who stand up in defence of the culture, language and political rights that are based on their ethnic identity." It is interesting to note that within such groups one can find young teachers whose education aimed at

V From Intellectual Dependence to Creativity

de-Indianising them and who now "affirm that the differences between Indians and non-Indians are legitimate and that the problems of Mexican society in general are attributable not to the existence of ethnic pluralism, but to the relations of domination to which [their] peoples have been subjected". They do not want the bilingual-bicultural teaching which they are doing to be a means towards 'assimilation', and they reject "aspirations for upward mobility which are formulated in individual rather than collective terms".

Similar processes are taking place all over Latin America, and "the shame of being an Indian, which characterised many young students in the last decades, is being replaced by an affirmation of ethnic identity". Even officially, bilingualism and biculturalism are now accepted; and, although most schools continue to function just as they always have done, the change in the official political dissourse is nevertheless important.

The emergence of a new Indian intelligentsia raises important questions. In the first place, it must be asked *why* at this moment there arises a new intelligentsia affirming its Indian identity. One reason is that the government intended the educational system to be a means of acculturation and assimilation, but the dominant society "offered structural and ideological resistances to the incorporation of a growing number of educated young people of Indian origin". In these conditions it was obvious that these peoples had to look for another alternative; and they found it in the defence and expansion of "an educational system exclusively serving the Indian population and remaining necessarily under its control". "On the other hand, at the international level, the political and national visibility of other ethnic and national minorities inspires the struggles of the Indians." Moreover, many people have come to realise that the social and economic problems of the Indian communities "not only have not been resolved by developmentalist and modernising projects", but have, in fact, "been continually aggravated" by them.

An especially important characteristic of the new Indian intelligentsia is that many of its members have been trained in the educational institutions of the dominant society; frequently they have both lost their Indian identity and then subsequently rediscovered it. "This process enabled them to gain a knowledge of the dominant society as well as a knowledge of their own from a perspective different from that normally acquired within the Indian communities. Some of them have training in the social sciences, but all of them have at least a direct experience of non-Indian society." They have set themselves the task of formulating an Indian world-view that is suitable for present-day realities. The political programme based on this view will be a plan for the transformation of the entire society; and "the new Indian intelligentsia of Latin America (like the native intelligentsias of Africa and Asia) faces the problem of taking over the social science developed in the West and simultaneously of transforming it critically in order to make possible what we can call their own endogenous development".

Here we encounter in practice an endogenous development of the social sciences. Coming quite close to positions developed by Dr Pandeya in the first session, Dr Bonfil Batalla explained this endogenous development theoretically as follows: "Social knowledge of social reality is in all cases based in particular experiences; it is historical, and it is precisely from [this] concrete nature [of social knowledge] that the ability to carry out specific forms of behaviour is derived. The social sciences which attempt to achieve universal validity must be incorporated into the [more widely diffused] social knowledge of society if they are going to effectively contribute to the generation of suitable and appropriate projects of development." It should also be stressed that endogenous development cannot be effected as an insular process, separated from the forms of knowledge created by other societies and particularly by those which exercise power and have achieved hegemony. "Not isolation, but incorporation of everything that is valid and useful into the fabric of genuine and dynamic knowledge will permit a real development of the social sciences placed at the service of an endogenous project."

Several measures will contribute to this endogenous development.

First of all, traditional social knowledge should be institutionalised, and "one obvious condition here is that the new institutions will have to be controlled by and made up of members of the group itself . . . .The process of institutionalisation is an indispensable prerequisite for constructing an endogenous base for the social sciences themselves, since it will permit systematising and formalising the traditional knowledge of socio-historical reality, legitimising it from both the inside and the outside, increasing it, disseminating it, incorporating it into the theoretical, methodological and factual baggage of the conventional social sciences, and reproducing it, through the training of specialists."

Secondly, there has to be a really *Indian dialogue* between traditional knowledge and present-day social science. This entails the recruitment of two types of specialists: those who are "carriers of traditional knowledge (who know the myths, legends and history of their people, who know their struggles or are specialists in traditional activities . . .) and those who have received training in the schools of the dominant society, but who have not lost – or who have recuperated – their Indian identity". Between these two there can be a real co-operation, the aim of which is "not to make traditional 'wise men' into 'informants'; but to recognise them as interlocutors and mentors of the 'new Indians' who are constructing the Indian knowledge and thought of today".

A third objective is the incorporation of this eventually institutionalised knowledge into development programmes – and not only into official programmes imposed from outside, but more importantly as the foundation for the management of internally generated projects. "Perhaps projects carried out along these lines can contribute to an ever more rigorous knowledge of social reality, in accordance with the necessities and historical perspectives of civilisation which every people possesses and which will guide their genuine development."

In a collaborative paper entitled *On the razor's edge*, Rector Pečujlić and Dr Vidaković argued that the present is a decisive moment in history. "Whether this epoch will represent a step towards the liberation of people and communities

V From Intellectual Dependence to Creativity

or a new technical barbarity, whether it will degenerate into lower or be reborn into higher forms of life, depends on [mankind's] capacity to offer a vision of a new world, a new civilisational alternative."

At present, "the hegemony of the old world is being maintained not only through repression, but also by means of cultural hegemony, the enslavement of consciousness — through the dominant patterns of production, technological and industrial development, patterns of consumption, types of urbanisation ...."

What, then, are the roles of science and technology at this crucial moment? It is clear that mankind now has at its command extremely powerful forces of production, "scientific and technical forces that no epoch of previous history could have envisaged". Automated systems make boring, repetitive labour unnecessary, and this should reduce the division between mental and physical labour. At the same time, new opportunities for collective responsibility and team-work arise, since "the complexity of the new technology makes the old hierarchical order inadequate for setting the new productive forces into motion".

In the field of education, "the classical industrial revolution created as its basis a type of elementary school which satisfied the need for a plain labour force . . . . But the technical transformations coming into existence now are connected with a cultural revolution of unprecedented proportions." Marx already foresaw this phenomenon when he said: "If industry develops, the creation of true wealth becomes less and less dependent on working hours and the amount of labour spent, and much more on the general state of science and technological progress . . . . The understanding of Nature, the development of human capabilities becomes the real pillar. The theft of the working time of others on which present-day wealth rests appears as a poor foundation compared with this newly developed one . . . . The free development of individuality and the reduction of labour time to a minimum suits this development . . . because [it is conducive to] the scientific and artistic education of man, which becomes possible due to free time and the resources which have been made." With the opportunities created by new developments, an ancient dream has come true: man *can* be liberated from the yoke of poverty, and the gap between rich and poor *can* be narrowed.

However, "like the ancient god Janus", the development of civilisation also has another face. The new sources of productive power have become destructive, both for Nature and for mankind: "Almost 20 per cent of [all] scientists are working on the discovery and application of means of world destruction." People constantly live at war or under the threat of war; hunger is taking a toll of twenty million lives each year, and Nature's resources are being robbed, wasted and polluted. The new productive powers, likewise, are contributing only to the wealth and power of a limited few. New devices are being created for manipulating and dehumanising the human soul: just look at the creation of artificial needs which foster the growth of nihilism, violence and neurosis. And the gap between rich and poor, far from being bridged, has grown tremendously: from 1:3 to 1:70. At a global level, the 'periphery' is undergoing a new kind of dependence: an external dependence supported by political force has been transformed into an internal, organic dependence enforced by the multinationals. If there is any unity in the world today, it is a unity "based on a relationship of fundamental inequality" – and it is nothing but "an old prejudice" to think that the roots of this situation lie "in a stubborn adherence of pre-capitalist structures". On the contrary, "the development of capitalism itself has given birth to underdevelopment in the past, just as it does today". Yet even the great majority of Marxists have not foreseen the dimensions of the present global crisis, and definitions of socialism to date do not take these dimensions adequately into account.

In these circumstances, the two faces of science and technology can be seen as an expression of the contradiction between two visions of the world. "The vision of the apocalypse is taking the place of the technological Utopia. Social progress was [formerly] equated with technical growth", but now a new slogan has appeared: 'Technology inevitably dehumanises and enslaves man.' It is of decisive importance, however, to realise that "science and technology are not negative powers *in themselves*". They only become so because of structures of social power and types of social organisation. Know-how by itself is "a means without an end, a mere potentiality . . . ; [and] what we need most of all is to turn this enormous potentiality into a new reality, to the benefit of the people". This means "breaking out of the encirclement of technological determinism which strips the world of everything that might suggest the capacities and actions of men", and which portrays technology as a new deity delivering commands from on high.

Indeed, our science and technology have not appeared in a vacuum. "The civilisation whose god-parents are profit-making and bureaucratic rule permeates them deeply"; and "such a deformed technology is no longer neutral – it has become an active factor which determines the attitudes of a producer towards his product, of a worker towards his labour, of an individual towards society, of man towards his environment. It becomes one of the foundations of the relationship of power, of the hierarchical division of society, ... a tool of domination over people and entire communities". Hence, adopting a society's technology also involves adopting some of its social structure, as can be seen from the experience of certain countries which have followed such a course of adoption in the early, difficult conditions of socialism with undeveloped productive forces. The dominant principles of profit, power and prestige have created a 'scientific subculture' with its own 'one-dimensional men' and its own separations between theory and practice, manual and intellectual work, professional knowledge and popular culture; such factors all work to prevent scientific and technological cadres from linking their knowledge into a broader context.

Nevertheless, "although science bears the stamp of the ruling civilisation in which it was born, it is never fully integrated into a system"; and therefore scientists often realise that they both "do belong and do not belong to the forces of social change". There is a definite desire for change within the world

V From Intellectual Dependence to Creativity

of science just as there is within the labour world. "However, the great changing of civilisation will not just emerge as the creation of 'technological prophets', as an automatic result of intellectual creation motivated by its own mysterious imperatives. It is, rather, a great social and cultural process in which the potentials of technology and science are put at the service of new goals, purposes and values, i.e. [at the service of] a different quality of human life. A new technology and new sources of energy will be born out of the new collective practice of this mass social movement. . . ." This practice will "by no means entail a rejection of the great accomplishments of science and technology"; but neither will it effect "a mere take-over of existing, ready-made parts out of which a new edifice can simply be erected".

Scientific and technological alternatives will have to be an integral part of a new historical project by which working people will accomplish their own emancipation; and "the decisive role in the formation of this new historical project is played" not by managers and technical experts who imagine that they alone know how to read the 'lessons' of history, but "by the *organic intelligentsia of the plebeian classes*", which is called to perform "a great elaboration of the collective aspirations . . . already developing within the plebeian masses". This organic intelligentsia differs from the "traditional satellite intelligentsia", which is purely imitative, because the former stands firmly on the foundations of its own national culture. Its members are not an élite, but the "representatives of their kin", "called upon to perform a revolutionary innovation of [their nation's] professions and [to create] new orientations in all fields of social life". As such, they are not just the travelling companions of the social movement but its key participants.

The transformation of the process of labour is becoming the decisive front of the social struggle, and the main questions in regard to it are those of *how* work is performed and of *what* is being produced. Developments in science and technology continue to revolutionise the production process every day; but new breakthroughs have liberated only a small minority from routine work, while "for a vast majority of workers it has brought forth new forms of monotonous routine labour . . . which, instead of being a source of health and sanity, becomes a punishment and curse". Decisions thus will have to be made about whether fragmentation of the labour process and disqualification of the workforce will continue and whether the organisation of labour will be transformed. People are, of course, already reacting in a number of ways against being made into robots. Most significantly, they are creating alternatives in which the technological organisation of labour is placed directly under the control of the producers themselves; and experience has shown that production actually increases when the workers have taken responsibility for the production process and done away with fragmented work. This shows that "the fragmentation of tasks is not simply a consequence of technology" but the result of technology being included in an oppressive system in which "the workers cannot be trusted" and in which a sharp division between manual and mental labour is reinforced as a typical means for domination. Isn't it indeed ironical that just as productive tasks are made more and more interchangeable and narrow, the hierarchy of non-productive functions becomes more and more oppressively baroque! It must be thoroughly understood that this division and domination is not so much necessary to ensure production as it is to ensure "the reproduction of capital and of bureaucratic (technocratic) power". One of the great sociological laws of modern society can be stated as follows: "The prestige carried by people in modern industrial society varies in inverse proportion to their closeness to actual production." This law must be overturned. One important means for doing so will be to reverse the tendency towards fragmentation of the labour process and disqualification of productive workers, for which, in any case, there is "no technical necessity". Nowadays "there is no need to condemn anyone to unqualified and stultifying tasks for an entire working life"; and the process of production can and should be organised in such a way that it will simultaneously function as a process of continuous education.

Another ploy of the ruling classes that must be abolished is that of widespread unemployment. Security of employment for the masses is now an essential precondition for the success of alternatives in both the developing and the developed worlds. A solution to unemployment in both 'worlds' might be found in simultaneously encouraging "*modern automated* production" and "*decentralised* production". "Quality products made in large series in an automated process of production would satisfy [basic] needs and would be [complemented] by a multitude of decentralised, local, self-initiated productive units ...." In the poorer countries, especially, emphasis should be placed on developing "production by the masses" (Gandhi) by mobilising their skills and "supporting them with first-class tools".

"The potentially deepest civilisational alternative is that embodied in the demand for essentially *shorter working hours*." Indeed, fixing shorter working hours for people and longer working hours for machines would be an excellent way of creating more jobs; it would also act as an incentive for inventing time-saving devices. Shorter working hours would contribute to overcoming the division of labour, facilitate a link between work and education and thereby enable every worker to constantly enrich his practical and theoretic knowledge.

In addition to such alternatives pertaining to the character and conditions of work, it is also necessary to critically reappraise which products should be produced and which needs satisfied. At present there is a deep crisis affecting modern modes of consumption, both individual and social. But what remains concealed is the systematic nature of this crisis – "the deep, organic relationship between the ecological crisis and the crisis in the mode of production". It is really "only a single logic which is reducing man to a commodity and destroying Nature ruthlessly. However, solutions do not lie in putting a stop to growth while preserving the same system, as proposed by the Club of Rome." What is really needed is "a great inversion": production should be geared towards what is necessary for all, and not towards enhancing privileges and hierarchies. There

V From Intellectual Dependence to Creativity

must also be a tendency towards increasing the quality and durability of articles, and towards more collective services.

Besides labour organisation and consumption patterns, a third aspect of life requiring change is that of living conditions. The development of cities is subject now to the logic of profit, real-estate speculation and city rents, all of which have transformed urban life into a monstrosity. Slums exist side by side with opulence; dwelling areas are destroyed to make way for flashy architectural colossuses; community services are becoming more and more expensive; and the quality of collective life is constantly being reduced. In the Third World, the major cities are surrounded by belts of incredible misery. People are drawn to the satellite industries, and this causes a breakdown in agriculture. In opposition to these trends, there are struggles for a new, alternative urban life. An example can be cited from the city of Bologna, where housing is being restored and rent kept cheap; public transport alone is allowed within the city, and that is free!

A fourth struggle is that for decent health care. Dehumanising working and living conditions take their toll in human lives and health, and yet medicine is oriented not to prevention but simply to repairing the human machine and returning it to working condition". Health is usually considered on a strictly individual basis; "like Shylock", the bosses are trying to get their pound of flesh "to the detriment of the living and working conditions" of the ordinary people. As Rector Pećujlić had observed during his intervention in the first session, scientific-technological innovation will only serve to improve these general conditions determining health *if* it is made the object of mass social action.

In conclusion, it can be said that science and technology will thoroughly serve human liberation only when based on the complementary unity of self-reliance and solidarity, of autonomy and a "new universality". "Intellectual creativity is only possible if the social ground is autonomous", and the 'concept of specificity' must encompass all dimensions of life, including "a strategy of technological development which is not restricted to the adoption of the patterns of others". This requires a search for a kind of development that will not dehumanise people and their environment, a development whose fruits will go primarily to the working people. Obviously this endogenously controlled development is not to be equated with confinement within one's boundaries; it is, rather, a necessary prerequisite for any mutually beneficial 'bridges'. Yet "without mutuality there can be no autonomy". We are striving towards a world that will be *one*, but one in which relations of hegemony will be replaced by "a *pluralism* of cultures". The liberation of the potentials of the entire world is the stake.

"Differences will remain. But the decisive question is whether they will lead to mutual complementariness or whether they will turn into hostility, antagonism." Moved by the great forces of national liberation and social revolution, the peoples of Asia, Africa and Latin America "are opening the gigantic hidden creativity-potentials of mankind, bearing the most valuable fruits in the creation of a new world". But, on the other hand, "the forces of hegemony are becoming the protagonists of the negative, antagonistic aspects of the integration of mankind", protagonists of "subordination" and the "annihilation of autonomy". It is crucial "to oppose this logic which disintegrates the unity of the working masses of the world; [it is crucial] *not* to enforce partial practice and truth as the only ones, not to present part of the sky as the entire horizon". In fact, "only autonomy, independence and equality can be a path leading towards a universal richness — towards a world enriched by the original and unrepeatable creativity of every civilisation. Deprived of this, interdependence is not a way to mutual enrichment, but an obstruction to the growth of civilisation."

Discussion

As the first speaker in the discussion, *Dr Wallerstein* commended Prof. Bonfil Batalla and Rector Pečujlić for reminding the conference that, besides the physical and biological sciences, the social sciences are also a part of science. He observed that most of the participants in the conference were, in fact, social scientists and that "even those who were trained as natural scientists or as engineers are speaking here in their capacity as social scientists; that is, this is a social scientific discussion of the role of science and technology in the transformation of the world. And therefore it behoves us to reflect somewhat consciously on the role of social science in the transformation of the world in the context of our discussions." It is good to emphasise endogenous intellectual creativity not only because it will give more power to presently non-powerful areas of the world and thus make for a more egalitarian world, but also because it contributes to the creation of a sounder, richer world culture.

Dr Wallerstein felt that Mr Blue had provided key categories in which to discuss the respective levels of science and technology in different cultures. Relating this to the social sciences, he said that it was not certain that Western social science has reached a higher stage than that of the rest of the world; "but in so far as it has, surely the 'fusion point' is far from having been reached: if it has not been reached in medicine, it certainly has not been reached in the social sciences, and . . . the point of this project is, in fact, exactly to facilitate arrival at the fusion point in the development of ecumenical social science".

Dr del Campo then took the floor to consider four points. He first of all raised several questions related to the fourth session and the introduction of the idea of power into the discussion of science and technology. One must consider both whether and precisely how the development of science and technology is affected by different *political* regimes, e.g. by the presence of a dictatorial or of a democratic regime. Do democratic decentralisation and the cost of democracy, in fact, hamper science and technology? Or, on the other hand, can science and technology foster and be fostered by democracy?

In turn, it is necessary to consider the sorts of *social* structure conducive to a scientific 'take-off'. It is clear that what has traditionally been considered 'European' science in fact has many roots elsewhere; and it is also recognised

V From Intellectual Dependence to Creativity 147

that in non-European cultures knowledge was often institutionalised and was thus really science and not just 'folklore'. It is not just a coincidence, however, that Europe was able to take over the achievements of non-European peoples and use them to bring about changes in a way that no other country or civilisation had done before. As a key to explaining this difference, "it should be remembered that the use of science by Europeans took place at a time of discoveries (i.e. at a time when the world was opening up) and also, . . . in a context of rebellion against authority". As Dr Nakaoka pointed out for the subsequent case of Japan, each 'leap' in European science took place in a particular historical context with a particular, historically determined social structure; and the social structure both made those leaps possible and responded to them.

This makes it pertinent to consider the role which the social sciences play in the introduction of science and technology into different societies today. Can science and technology be introduced into any society, irrespective of that society's own system of social science? In many different parts of the world, attempts are being made to master the natural sciences and technology, while at the same time rejecting "as ideology what we call 'social science'". It is therefore necessary to define exactly what is meant by 'social science' and to clarify whether this is necessarily going to be based on the social sciences developed in Europe during the nineteenth century.

Finally, Dr del Campo called the conference's attention to the special position of countries such as Spain which, strictly speaking, belong neither to the periphery nor to the core of the contemporary world economy. Such countries are "not systematically developed". They may have great literary or other traditions, and they may be specialised in certain economic sectors, but they are still dependent from the point of view of science and technology.

Dr Imré Marton followed next and spoke about the social vocation of intellectuals. Dr Marton said that the world today has too many graduates and too few real intellectuals. Democratisation in education has opened up middle and higher education to more people, and the general cultural level has been raised; but the students are not being trained as responsible intellectuals: their horizons remain fixed especially on 'getting their diplomas' and not on becoming really qualified or on diligently cultivating a critical attitude towards their societies and towards themselves. In turn, they lack creativity. At the same time, despite the increasing number of students, the contradiction between manual and intellectual labour has sharpened. "There is not only disqualification of manual labour, but also a depreciation of it." This phenomenon is not only typical of the developed world; it can, for instance, also be found in Africa. It can even be said that in conditions of underdevelopment an increase in the rate of scholarisation presents the danger of increasing the non-productive urban population. University graduates, in particular, often enter 'automatically' into parasitic government positions, relying on tribal support for their ambitions and thereby distorting not only "primitive ethnic solidarities", but also the formation of classes and nations.

According to Dr Marton, scholarisation should be extended at all levels, but it is necessary to keep the goals of educational training in sight and to establish a correlation between such training and the possibilities for using it. At present, increases in numbers of students are not linked to the requirements of economic growth. The imbalance is, of course, a function of the general dominance exercised by the Centre over the periphery, and, therefore, an adequate strategy for any genuine progress requires a global framework free from either Eurocentrism or *tiermondisme* and able to generate alternatives based on the accumulated experience of all three 'worlds'.

Dr Alexander Kwapong then intervened to make three observations. He said first of all that he had been struck at the recent UNCSTD conference in Vienna by the "intractable gap between science and technology, on the one hand, and the political and social processes, on the other". "Everybody agreed that one of the disappointments of UNCSTD was its failure to bring statesmen and government representatives together effectively with the scientists and technologists and to effect the necessary interaction. . . . " Indeed, effecting such interaction does seem to be one of the most important challenges now facing many nations in the world; and it is necessary, therefore, to "define and analyse the type of forum in which the two sides can be brought effectively together and made to interact".

Dr Kwapong's second point concerned the necessity of understanding the historical dimension in development. He considered Japanese history, for example, of special interest to Third World countries; and he especially admired Tokugawa Ieyasu (d.+1616) for having founded a dynasty which upheld Japan's national sovereignty and preserved her native culture for two centuries before the Meiji Restoration opened the way for modernisation. In general, it is important for Third World countries to constantly emphasise the historical dimension, because "development is a long process of gestation" in which exogenous and endogenous factors are always interacting. And Dr Damjanović was quite right in pointing out that real development consists not just in mastering particular technologies, but especially in creating the human skills that will be able to generate new scientific-technological innovations. Thus, one should be aware of the special importance which should be given to education, and an infrastructure of human experience must be built.

Finally, the knowledge conveyed at conferences such as the present one must be disseminated and applied to practical realities, such as answering the question of how to overcome "the all-pervading problem of corruption in national life". Even with all the best analyses of alternatives available, there will be no genuine development if politicians "take decisions which are . . . completely 'exogenous' to the real needs of their peoples, but very 'endogenous' to their own pockets and numbered accounts in industrialised countries". It is necessary to concentrate on dealing with this "hidden agenda which is frustrating real development"; and attention should be focused on moving from theory to practice.

V From Intellectual Dependence to Creativity

Dr Pandeya then observed that a point common to all five position papers had been that, in general, "science and technology . . . become a force for transformation only when they are organically linked up . . . with the total cultural resources of a community". This "fusion" or "confluence" is the crucial prerequisite for converting science and technology from an instrument of oppression into a resource for building a new future.

In studies concerning the scientific-technological development of Third World countries, there are often "intellectual blinders" which quite effectively prevent one from seeing things correctly. On this score, the dominant academic stereotype is, perhaps, best formulated in "that famous ethnocentric question" raised by Max Weber at the beginning of the century, viz. why was it that modern science as well as the spirit of capitalism arose only in Europe? According to Dr Pandeya, this way of posing the question is based on a "false" approach to the problem; and Dr Damjanović had provided the "long-awaited corrective" when he put forward "the contrary position . . . that no society, primitive or overdeveloped, can reproduce and sustain itself as a society unless it has developed the capacity to generate the minimum knowledge-base necessary for its functioning and meaningful living; and one of the components of this social knowledge-base has got to be the scientific-technological one". If a society does not stress this component "in the lop-sided manner" which European civilisation has done since the Renaissance, this is not necessarily because it lacks the capacity to do so, but because its "total capacity for generating knowledge" has been distributed differently from that of modern Europe. The generation and classification of knowledge (and so of science and technology) must be recognised as universal premises for all societies; and all countries are confronted with the necessity of creating and maintaining their capacity for generating knowledge. Thus, scientific-technological development must be seen in terms of the dynamics of "the stagnation or the revival of societies"; and if scientific and technological development is to take place in Third World countries, these dynamics must be tapped in order to "give rise to a leap of reclassification and re-emphasis in the basic social populace" and to generate a revitalised knowledge-base with a new "social, egalitarian purpose".

Finally, Dr Pandeya recalled that 28 per cent of the population of India is made up of tribal minorities, who economically and politically are among the most deprived groups in the entire country. In the name of preserving their cultures from the ravages of modern society, certain government programmes have at times been inaugurated which, in practice, work to turn these peoples into museum pieces. At other times, other projects were undertaken which aimed rather at remaking them in the image of modern capitalist societies. Contrary to both of these approaches, however, what is necessary is for such people themselves to consciously take their place in the transformation of the world; they certainly do not need to be confined in a mould.

In the next intervention, Dr Pinguelli Rosa noted that, although science and technology had been treated together as a unit throughout most of the conference, they are frequently distinguished quite sharply in reality. For example, as pointed out by Dr Leite Lopes, many underdeveloped countries have already devoted large investments to developing science; but, because economic and political conditions in such countries lead enterprises to buy the technologies they need from the multinational corporations, the scientific institutions which are set up remain isolated from technological research and development that might be advantageously utilised for transforming the structure of production in these countries. The internationalisation of science and technology which is at present taking place is certainly one that is being imposed by the multinational corporations; but this internationalisation must be considered in detail, and it is not sufficient merely to take it as "an immutable input data".

Dr Pinguelli Rosa also noted that throughout the conference there had been an ambivalence towards science and technology: at one moment they had been treated as good things which had to be imported for the national welfare; at the next, they were considered bad things which brought about harmful effects, socially and otherwise. In fact, "the underdeveloped countries sometimes . . . have to start from things that rich countries have established"; but this is no reason to remain passive when confronted by the policies of the multinational corporations. Instead, it is necessary to break the hold of the multinationals and to devise a new sort of internationalisation of science and technology.

In a short comment on Dr Pinguelli Rosa's intervention, *Dr Abdel-Malek* said that simplistic assertions which label science and technology as good or bad in themselves amount to a denial of the role which politics play in the world. Yet it is precisely because of the political dimension that different countries and different sorts of countries have "conflicting and often antagonistic priorities", in relation, among other things, to scientific-technological development. For this reason peoples in the developing countries must reject the various forms of interest-laden obscurantism advanced in relation to science and technology, and they must define their own goals and priorities themselves. Now, at present, "in major countries of the so-called 'South', the forces of the popular movements, the peasants, workers and intelligentsia, are running frontally against the crescendo of scientism in the West. . . . At the same time, they are being made aware that the rush towards development . . . will allow itself to be collared in an inevitable deadlock unless a different civilisational pattern can be proposed." According to Dr Abdel-Malek, the project of defining such a pattern is a matter neither of evading political reality nor of cultural escapism, but rather "of linking power and culture". It requires not utopianism but realistic fraternal visions.

In the last intervention of the evening, *Dr Furtado* called the attention of the conference to "the danger of certain concepts, such as that of ecumenical science". Observing that the fundamental problem facing the conference was the relationship between science and civilisation, Dr Furtado warned against the illusion of believing that science as it is practised at any particular moment might be the only possible science, fully comprehensive and exhaustive. The problems

V From Intellectual Dependence to Creativity

with which science deals at any point in time are conditioned by particular social relations, ruling groups, power systems and financial resources; and thus it is necessary to keep in mind that science may be "more ecumenical to one particular civilisation" which is dominant than to others which are "striving to survive".

Appendix 1
Reports on Sections and General Report on the International Seminar

The orientations, as well as tonality, of the discussions dealing with each of the four main themes of the International Seminar were summarised in the four Reports on each of the four themes, as submitted for the final plenary discussion.

In his report on the first theme, *Science and technology as formative factors of contemporary civilisation: from domination to liberation*, Dr James A. Maraj thus summarised the "sharpening of differences which had emerged" and the positions arrived at:

"(a) There is a need for a much more vigorous examination of the relationship between the way a technology is applied, the technology itself and the basic science from which it grew.

"(b) In pursuing the examination, attention should be focused on the role of technology as a factor in the social, economic and cultural aspects taken individually and cumulatively. The interaction between the various aspects should also be closely observed as well as such matters as: Is the problem one of technology itself or of its management, or of the resource base? etc.

"(c) It is particularly important to re-examine these relationships as we are entering a new era following the current world crisis.

"(d) Technology is not an end in itself. It has been preconditioned by social goals and these goals need to be clearly articulated. In particular, the human factor has to be emphasised, not only from the standpoint of the individual, but also from that of the social group, as we strive towards egalitarianism in terms of equality of opportunity. The search for a fraternal, convivial society should recognise both cultural identity and diversity.

"(e) In accepting the proposition that science and technology are socially conditioned, it was thought that it would be useful to study the application of

Appendix 1

various specific technologies to determine whether the goals being pursued are in fact being achieved or whether the technologies imposed by wrong motivations themselves alter the character and the nature of the technology.

"(f) While recognising that various dilemmas would have to be confronted, it was thought that some of these could be made less difficult to cope with, if clearly defined social criteria could be stated and adequate methods agreed upon for assessing and forecasting technologies.

"(g) It was noted that while science has to a large extent been decentralised, this is not so with technological development. The latter is still heavily monopolised by a few powers.

"(h) The link between technological development proceeding from a scientific base was seriously questioned, and several reasons were given for the non-automatic emergence of technological development even where a strong science base existed.

"(i) It was concluded that the various parameters of the social field had to be carefully examined before deciding on technological development or adaptation and that the entire social system itself would be the determinant of the extent to which technological development would succeed in effective transformation."

Dr Cuthbert K. Omari, in his report on theme two, *Technology generation and transfer: transformation alternatives*, noted the elaboration of the following points:

"In relation to the transfer of technology from developed to developing countries it was observed that there is a problem of language. Usually the technological tool or technique developed in western countries has a functional role. There is meaning attached to it. When it is transferred to the developing countries whose language is not the original language in which the technology was invented, it becomes a problem, for the people using it will have no relation to the original meaning of the name given to the tool or the part of the machine. Always the language of invention has a symbolic meaning. This problem of imitation and appropriation was further elaborated by giving examples of African experiences.

(a) Most countries have different languages within one country; but, if there is one language, transfer may be easier.
(b) The problem of under-population was also noted; this is a problem in relation to the mobilisation of productive forces and the marketing system.

"It was noted, however, that a country whose traditional technology had reached a certain level might adopt a foreign technology with fewer problems.

"In relation to Africa a question was raised as to what extent an African can become modern without losing his/her identity. How one can remain in the past, tradition, without bringing about stagnation in social development? This point was not discussed fully.

"It was also pointed out that we are witnessing universalism in our days. People share the same cultures, but this again has its perils. It may bring about conflicts and endanger the survival of mankind.

"The problem of self-reliance in relation to transfer of technology was discussed. It was pointed out that it is impossible to resist science and technology in developing countries, but how can we adopt them without being dominated by the developed countries?

"The suggestion was made that the communication system among developing countries should be strengthened. This will help to control information, and it is within the area of collective self-reliance in the Third World. From there, then, information can go to the developed countries. This may further help to prevent the side-effects of imitation of science and technology.

"The problem is how to change from the interdependency of domination to mutual interdependency."

Reporting on theme three, *Biology, medicine and the future of mankind*, Mr Gregory Blue dealt with several themes which emerged through the discussion:

"(1) It was strongly questioned whether the attempt to apply a bio-sociological approach to contemporary human problems is valid. There was divergence about whether such an approach leads to illusions about reconciling irreconcilable forces in the political sphere, or whether it can aid in constructing a non-confrontational model for dealing with problems.

"It was pointed out that the distortion of social time-sensibility is an extremely great problem, which is leading to a generation gap on a world scale. Productivism tends to destroy people's collective memories and hence to make impossible a collective present.

"(2) It was pointed out that transformation of existing medical relations must involve a clear understanding of the role of the transnational pharmaceutical companies. It was also said that physicians must choose between serving these companies or serving the real needs of the people.

"(3) In regard to medical practice, it is necessary to keep in mind the mutual interaction between the individual human organism and the external environment. Many types of disease are impossible to cure without a previous transformation of external physical or social conditions. In such a situation, it is extremely misleading to think that medical success can be attained at the level of the individual alone.

"(4) It is incorrect to consider science and technology as panaceas for all problems, as independent and isolated from other factors in the world. It was observed that the majority of technological advances since 1945 have been consequences of the armaments race and that this race now dominates technological advance. It is, thus, true that science and technology are sources of power in the present transformation of the world, but they must be viewed within the full context of global power relations.

"(5) It was observed that, since 1947, the greatest overall growth in the world has taken place in the underdeveloped countries, and that this growth has for the

Appendix 1

most part occurred in industrial production, whereas growth in the developed world has mostly occurred in the service sectors.

"(6) It was said that the negative effects of industrial techniques are only inevitable so long as there is no social control over them. Also, it is necessary to take into account all aspects of transformation planning: e.g., it is absurd to promote schemes for 'rural development' which ignore subsequent urban overcrowding, etc."

Reporting on theme four, *The control of space and power*, Dr Vladimir Štambuk presented the following systematic overview:

"The largest part of the discussion was concentrated on the question of power and its role in the transformation of society, including technology and science. Five elements were suggested at the beginning as the sources of power: control of international markets; control of international finance; control of non-renewable resources; control of cheap manpower; and control of technology. It was stressed that developing countries can control the first four factors and are yet unable to control the fifth. To these elements, others were added during the discussion. Control of knowledge and information was stressed as the most important aspect through which influence is being realised over the developing countries.

"Another element was added: political power; and a criticism was expressed concerning the control which the developing countries have over the first four elements. Expressed opinion was judged as over-optimistic.

"The role of politicians was also discussed and the ambiguity of their position between the pressure of the masses for solving everyday needs and the impossibility of leading political structures in developing countries to solve them accordingly. That is why politicians make deals with multinational companies and develop industrialised societies, in order to maintain their power. The multinationals, using the principle of divide and rule, augment their domination over developing countries by a process in which political structures have often been included. There is a feeling that politicians should be educated, too. This education should be concentrated on the political strategies open to developing countries. Scientists and university people are not always welcome as advisers to politicians, because they usually put forward views which are not in accordance with the choices open at the pragmatic political level. In that context, the role of the United Nations University, as an objective international institution, was stressed, and a hope was expressed that the activity of the University will be more intensive in that direction.

"It seems that there are three possible relations in the promotion of science and technology between highly developed countries and multinationals, on one side, and the developing countries, on the other side. They are:

(a) developing countries becoming client states under the political hegemony of an industrial state, which facilitates the operations of transnational corporations;

(b) such countries becoming dependent on the transnational corporations, under the influence of the World Bank and the International Monetary Fund; and
(c) such countries developing regional co-operation as emphasised in the document of the non-aligned countries, and discussed at the seminar as collective self-reliance.

The third relation is the most acceptable one, but has not produced the expected results. Countries which have opted for the first relation register economic growth without effecting much-needed social-cultural transformation.

"In the dichotomy of the hegemony of economism and the hegomony of ethical normativism, we have to look for a solution which will be related to the parameters of power. The problem is: How to undergo transformation but remain sovereign and creative?

"In this aspect, the role of the State is very important. The topic of the seminar was not related to problems of the State, but its role must be emphasised in further discussions. Thinking about 'science and technology', and the transformation of societies, we must insist on reality and the possibilities which reality is offering us.

"The disarray of the present world situation seems to offer – in spite of its being fraught with obvious dangers – a wider variety of options for the developing countries to establish better control over their future development. While admitting the existence of possibilities for transformations in social, economic, and political structures, it should also be noted that ensuing conflicts will be more complex and sharp and that the capabilities of the adversaries are much greater, qualitatively and quantitatively.

"On one level, we see that in the long run it is the developing countries which have expanding markets, financial resources, non-renewable resources, and manpower reserves. They do not, as yet, command technology as a resource which might make up for deficiency in any of the other resources, and technology has become the main source of power – a fact that highlights the role of science and technology in world transformation.

"On a more profound level, it is recognised that no one set of variables can be operationalised without addressing the specificities of the various particular situations of which there is considerable variety in the world today. The potentials amenable to mobilisation in a situation where a nation has a long history of consolidated existence are quite different from those where even the concept of nationhood is new or inapplicable.

"The specificities should, furthermore, be coupled with other factors in the international situation. There are emerging nowadays in the developed world allies of the developing world, particularly in the area of the production and dissemination of knowledge. The UN University can play an important role here, one that may create a consciousness that will later on trigger significant results.

Appendix 1

Public opinion in the North is gradually mobilising against intervention through direct action in Third World affairs.

"Attempts at regaining social control of techno-economic activities are thwarted in the name of economic rationality, efficiency and the adoption of consumption patterns that favour the expansion of the activities of TNCs in alliance with local capital and even state enterprises.

"The importance of specificity can also be seen in the isolation of the scientific-technological potentials of nations and states and in the fact that political constraints prevent their fusion into critical forces that would be effective in transformation.

"Contrary to options of isolation or of becoming a vassal or client state dependent on TNCs, the option of regional joint action needs exploration within this framework; and emphasis should be placed on complementarities, leading to greater national control and power, within the existing constraints. These aspects have been examined in considerable detail in the many documents proposed by the UNCTAD Conferences, reflecting the genuine concerns and requirements of the newly liberated, socio-economically backward countries of the developing world. The only way out is to generate simultaneously social mobilisation for cohesive socio-economic transformation in each specific country, together with linking efforts for joint concerted action based on collective self-reliance – the strategy spelt out in the political and economic documents of the summit conferences of the non-aligned nations from Algeria to Colombo and Havana.

"It has been stressed also that a new source of power is assuming an increasing importance: the control of information and knowledge. The international mass media are diffusing a world culture based on the ideology and system of values of the industrialised countries. 65 per cent of information messages are now produced in and diffused from the United States. The press, radio and TV are such powerful instruments that they are able not only to manipulate public opinion but also, as has occurred, to destabilise governments. The world information system is now an ideological apparatus which contributes to the continuance of the existing international order.

"Special emphasis was given to scientific knowledge and technical information as important tools for the control of power in the world. Such a kind of control has often been used by the rich countries with the goal of maintaining their domination over the underdeveloped countries. In this sense, the role of multinational corporations is precisely that of controlling productive activities in underdeveloped countries which have no autonomy to decide their own future.

"The optimistic point of view that multinational corporations play some positive role in broadening modern technology is largely negated by the effect of domination of underdeveloped countries in all aspects: economic, cultural, etc.

"Science and technology related to the concept of development are emerging as a new ideology of modernisation. This is part of an ideological attempt to control the development of developing countries and is obscuring the real relations between developed and developing countries. The causal relations between development and underdevelopment are lost, and causes of underdevelopment are hidden from view. This is an attempt to eliminate the values of socialist revolution with the ideological concepts of modernisation."

Finally, Dr Ahmad Yousef Hassan, in his report on theme five, *From intellectual dependence to creativity*, wrote:

"Science and technology are not the products of modern societies alone. Science and technology in Europe have been influenced by and based upon the heritage of other civilisations. Although distinctively modern science has arisen in Europe since the sixteenth century, the birth of modern science owed much to the preceding achievements of Greek, Indian, Chinese, and Islamic/Arabic science. For example, in the first fifteen centuries of our era, a large number of mechanical and other innovations were transmitted from China to Europe, and these were large factors not only in revolutionising mediaeval Europe, but also in the constitution of modern science itself. The same can be said of the transmission of techniques and the exact sciences from the Arabic world to the Latin West through commercial and cultural contact and through the movement of translations into Latin between the ninth and fourteenth centuries. This point shows that science and technology are not creations of any one civilisation alone. Science can flourish because of beneficial social and political conditions. When we consider the birth of modern science we see that it was also linked to the particular social evolution that took place in the West.

"Science and technology continued to expand and flourish in the West, whereas they developed at a much slower rate in the civilisations of the East. Science and technology as developed in Europe consequently were employed as means for domination and suppression, which tended to hinder similar developments within the other regions of the world.

"If favourable conditions can be created for the development of science and technology in any country, whether in the East or in the West, then there will be a flourishing, as can be seen from the Japanese experience. Here we have the example of a 'late-coming' Asian country which was underdeveloped in science and technology, yet favourable conditions were created in that country because of the lack of foreign domination, because of the presence of an independent, strong central government, and because of the interaction of endogenous and exogenous factors. The Japanese experience shows that modern technology can be successfully incorporated into local culture, provided that certain prerequisites are met.

"It seems beyond doubt that human freedom and national liberty depend on economic independence. Having in mind the role which science and technology play in economic development, attention must be given to the discrepancy in the levels of research and education between the 'North' and the 'South', that is, between developed nations leading in science, and the bulk of humanity, still striving in poverty for cultural recognition and freedom from domination. But

Appendix 1

pleading for more science and technology alone does not seem to very much affect cultural emancipation, the dominant aim of which is to preserve and revive national roots and culture and so to open the perspective of human civilisation as a plurality of national cultures. As was outlined above, throughout history science and technology have been deeply rooted in the human race, and all attempts to ascribe them to any one nation, or group, as local achievements or characteristics, are false. Whatever the national culture is like, science and technology can fit in, as complements. And though it may seem a paradox, no national culture will survive, unless it makes space within itself for the scientific-technological culture.

"Science cannot be developed primarily through the needs of local, divided practical activities. It is also impossible to plan scientific application in detail. Therefore, a broad population should be cultivated in science in order to help society to develop in a competent way. Similarly, competence in technology is not a matter of choice or of some local priorities. It is part of the basic culture of a broad population.

"Science and technology are a way of thinking. They deal with basic things in the human environment and in humans themselves. Therefore it is not possible for a society to benefit from science and technology without being exposed to their influence on human behaviour.

"Far from looking upon science and technology as creating unemployment, they must be considered as liberating man from dull work and over-work. Science and technology are thus prerequisites of emancipation and development.

"When speaking about science and technology it was usually implied that we were speaking only of physical and biological sciences. The role of social sciences was usually ignored. All societies need scientific social knowledge in order to build their own futures. Western social sciences do not have universal validity, and they cannot adequately cope with the problems faced by other peoples with different civilisations. The knowledge of the members of certain cultures about their own societies is not institutionalised or systematically organised. The social science which is needed to help these peoples in their liberation effort must include both the systematic knowledge of their own societies and of the dominant hegemonic societies.

"Hegemony is not being maintained only through repression, but also through cultural domination. The ability to conceive new visions is becoming decisive. We are confronted by the greatest challenge of all, the creation of a knowledge that is suited to our epoch. There are two faces to science and technology. There is the vision of social and economic growth, and there is the vision of an uncertain future and the illusory criticism of technology which gave rise to the slogan 'protect us from technology'. However, it is of decisive importance to realise that science and technology are not negative powers in themselves; they turn into that when becoming part of an antagonistic social arrangement. What we are seeking is a new type of society or civilisation which is to be a more favourable framework for the development of the authentic potentials of man.

"The new culture or civilisation cannot be built without international solidarity. Without mutuality, there is no autonomy.

"The coming era opens a glorious but also critical period of overall interdependence. We are living in a planetary world society. A pluralism of cultures is necessary in order to have the world become a society which is not uniform and indistinguishable. Only autonomy, independence, and equality can lead toward universal richness. Differences will remain. But the decisive question is whether they will lead to a mutual complementarity, or whether they will turn into hostility and antagonism."

It remains for the General Report on the International Seminar, co-authored by Drs Tsurumi Kazuko, Rajko Tomović and A. N. Pandeya, the joint General Rapporteurs, to sum up the emerging convergence and, more so, the emerging problématique:

"(1) The first international seminar, dedicated to the investigation of one of the crucial items on the agenda of our age — the role of science and technology in the transformation of the world — met in a context of expectations, clearly articulated by the Project Co-ordinator, Dr Anouar Abdel-Malek, and the Conference Chairman, Dr Miroslav Pečujlić, Rector of the University of Belgrade, in the capital city of the Federal Socialist Republic of Yugoslavia, which vigorously proceeds along the paths of constructive mediation between the different spheres in the world of power and culture at work in our times. The central character of our times, of the real world in our times, is implicit in the transformation of all the dimensions of the life of human societies — a transformation which is neither unilinear nor synchronic, but involves the different sectors of social life and activity — economic production, patterns of power, social cohesiveness, cultural identity, civilisational projects, political ideologies, religious formations, philosophies, myths, etc., covering the entire span of the infrastructure and superstructure of society. The question arises: how can this transformation of the world be related to the social and human sciences, political and social theories, the philosophical quest for humane vision — in short, the cultural and civilisational dimensions of our life tomorrow — through structural modifications, through the remodelling in depth of the world as we know it today? And the general focus, within which such interrogations and deliberations as are relevant to the problématique could unfold themselves, yielding significant, converging insights, must inevitably couple science and technology with culture, culture/civilisation with power, in the belief that such confluences should become the meeting point of scholars and policy-makers; of specialists in the natural, mathematical, material, engineering, and life sciences with analysts and theoreticians of the sciences of man and society, of humanistic cultures and civilisational totalities. And the problématique, in all its complex ramifications, must continuously remain grounded in the firm territory of the crisis confronting us all — in the monstrous asymmetries of economic, political, scientific-technological, cultural/civilisational, informational/

Appendix 1 161

communicational resources, characterising the present distribution across the globe.

"(2) This complex problématique was explored in its major facets, comprising constellations of specific questions and issues, through five plenary sessions, focusing successively on Science and Technology as Formative Factors of Contemporary Civilisation; Technology Generation and Transfer: Transformation Alternatives; Biology, Medicine and the Future of Mankind; The Control of Space and Power; and From Intellectual Dependence to Creativity. The expositions, discussions, debates, interrogations, illustrative concretisations, insightful suggestions, reflections and observations — all the diverse forms of cognitive, exploratory activities that were triggered off by the earnest engagement of leading minds from the major cultural, socio-political zones of the globe — eventually took identifiable flow-patterns: mutually complementary, occasionaly converging, often ranged in debate-prone tendencies, sometimes in polar opposition, reflecting the real contradictions and divisions of our real-life situations. But, on balance, as the dialogue built up, gathered material and dynamism in its movement from the plenary session to the dialectical stage of in-depth reflections in the working sessions, it was impossible to escape the feeling of a general focusing slowly taking shape, of a broad convergence gradually unfolding as insights and thoughts started falling into place; of an overall deepening, extending, of our understanding; of the centrality of certain issues; and of the awareness that what had actually happened was a cognitive transformation that had overtaken us all!

"(3) As concluding reflections on the problématique, it must be advisable to take note of those areas where, relatively speaking, the shared insights and cognitive convergences appeared to be pronounced. Science, in its totality of domains — natural, human-social, cultural/civilisational — and technology were everywhere firmly and deeply embodied in the socio-political structures which determined their dominating/liberating functions. Their hidden social relations and hidden power-base, therefore, needed total transformation, if these resources were to be converted into a massive cultural/civilisational force for re-forming the greater part of the human societies into a more humane, democratic, just, and egalitarian future. The cultural question, then, was how to disseminate scientific insights to the people at large; how to integrate the dissociated sectors of science/technology with the foundational sectors of political-social policy formation and decision making; how to strengthen complementarities across differentiated orientations; how to identify and strengthen solidarity among humane, transforming, progressive sectors of humanity distributed across the existing divisions of socio-political boundaries; how to sharpen focus on the gearbox of changing, challenging priorities; how to cope with the ever-increasing pressures which hegemonistic, dominating centres were busy releasing at an exponential pace; how to mobilise and organise the vast, latent reserves of endogenous creativity of the vast majority of mankind for initiating, sustaining, and completing the transformations that are overdue, that admit of no procrastination, divergence, or masking. In this realm of confluence, where reflective activity suggested urgent action, we note the final thrust of the seminar deliberations."

Thus ended the First International Seminar of the United Nations University's Sub-project on 'The Transformation of the World' – itself part of the UNU Project 'Socio-cultural Development Alternatives in a Changing World (SCA)' – an innovative attempt to explore the different dimensions of the new international order, at the very heart of the preoccupations of the United Nations and world public opinion, at the initiative of the United Nations University, alongside the vision of U Thant.

In place of the usual contrapositions of infrastructure and superstructure, of the prevailing economistic approach to the problématique of the world order and its required transformation, this first major meeting centered, substantively and specifically, on the *transformational* processes – their nature, inner dialectics, external parameters, actors and forces involved on all sides – as a *unified set of interwoven circles*, whose exploration could, alone, give meaning to the contradictions and convergence, deeply ingrained differences and complementarity, and, perhaps more so, to the disconcerting and disconnectedness of tempi, deeply rooted in the objective societal conditions of different units of analysis and action, as well as in the different visions of the world at work in our times.

Four other International Seminars, following after this one held in Belgrade, will be devoted to other dimensions of the transformation of the world: Economy and Society; Culture and Thought; Religion and Philosophy; and, finally, The Making of the New International Order.

Appendix 2
Participants

Board of Honour

Prof. Dr Pavle Savić, President of the Serbian Academy of Science and Art
Prof. Dr Anton Vratusa, President of the Executive Council of SR Slovenia
Dr Milojko Drulović, Executive Secretary of the Presidency of CK SKJ
Dr Krsto Bulajić, Director of Federal Administration for Educational, Scientific, Cultural and Technical Cooperation

Organisation Committee

Chairman: Dr Miroslav Pečujlić, Rector of the University of Belgrade
Secretary: Dragisa Stijović, Director of the Department of International Relations, University of Belgrade

Members

Dr Rajko Tomović, Professor, Faculty of Electrical Engineering
Dr Zvonimir Damjanović, Professor and Manager of the Centre for Multidisciplined Sciences
Dr Jordan Pop-Jordanov, Professor, Faculty of Electrical Engineering
Dr Ljubisa Rakić, Professor, Faculty of Medicine
Dr Vlastimir Novaković, Dean of the Faculty of Mechanical Engineering
Dr Ines Wesley Tanasković, Chairman of the Council of the UN University
Dr Slobodan Ristić, Director of the Administration for International Scientific, Educational and Technical Co-operation of the SR of Serbia
Dr Zoran Vidaković, Professor, Faculty of Political Sciences, University of Sarajevo, and Faculty of Law, University of Belgrade
Dr Milos Nikolić, Member of the Centre for Social Investigations of the Central Committee Board of Yugoslav League of Communists
Dr Vladimir Štambuk, Professor, Faculty of Political Sciences
Spasoje Grdinic, Chairman of the Culture Centre of the University of Belgrade
Jelka Brajovic, Collaborator from the Federal Council of the SFR of Yugoslavia

Participants

Dr Radoslav Andjus, Member of the Serbian Academy of Sciences and Arts; Professor, Faculty of Natural Sciences, University of Belgrade

Dr Yves Barel, Director, Centre d'études des pratiques et representations des changements socio-économiques (CEPRES), Institut de prospective et de la politique de la science, Université de Grenoble, France

Dr Guillermo Bonfil Batalla, Director, Centro de Investigationes Superior, Instituto Nacional de Antropología e Historia, Cordoba, Mexico 7, D.F.

Dr Vesna Besarović, Professor, Faculty of Law, University of Belgrade

Mr Gregory Blue, Assistant to Dr Joseph Needham, East Asian History of Science Library, Cambridge, England

Dr Krsto Bulajić, Member of the Honorary Board of the Seminar; Director, Federal Administration for International Scientific, Educational, Cultural and Technical Co-operation, Belgrade

Dr Salustiano del Campo Urbano, Decano, Facultad de Ciencias Políticas y Sociología, Universidad Complutense de Madrid, Madrid, Spain

Dr Zvonimir Damjanović, Manager, Centre for Multidisciplined Sciences, Belgrade

Dr Aleksandar Despić, Member of the Serbian Academy of Sciences and Arts; Professor, Faculty of Technology and Metallurgy, University of Belgrade

Dr Aleksandar Djokić, Vice-Rector, University of Belgrade

Dr Milojko Drulović, Member of the Honorary Board of the Seminar; Executive Secretary of the Presidency of the Central Committee of the League of Communists of Yugoslavia, Belgrade

Dr Osama A. el-Kholy, Faculty of Engineering, University of Cairo, Egypt

Dr Celso Furtado, Professor, Faculty of Economics, University of Paris, Paris, France

Dr Ahmad Yousef Hassan, Rector, University of Aleppo, Syria

Dr Takeshi Hayashi, Councillor, Institute of Developing Economies, Tokyo, Japan

Dr Stuart Holland, MP, London, England

Dr Dusan Jaksić, Rector, University of Novi Sad, Novi Sad, Yugoslavia

Mr Ilija Janković, Head of the Centre for Transfer of Technology, Federal Administration for International Scientific, Cultural, Educational and Technical Co-operation, Belgrade

Dr Kenji Kawano, Director, Research Institute for Humanistic Studies, Kyoto University, Japan

Mr Miroslav Kis, Student Vice-Rector, University of Belgrade

Dr Everett Kleinjans, Chancellor, East-West Center, Honolulu, Hawaii, USA

Dr Henri Lefebvre, Paris, France

Dr Le Thàn Khôi, Professor of Sciences of Education, University of Paris 1, Paris, France

Dr Miloš Macura, Professor, University of Belgrade

Appendix 2

Prof. James A. Maraj, Vice-Chancellor, The University of the South Pacific, Suva, Fiji
Dr Imré Marton, Professor of Philosophy, Karl Marx University, Budapest, Hungary
Dr José A. Silva Michelena, Director, CENDES, Universidad Central de Venezuela, Caracas, Venezuela
Dr Vaso Milincevic, Secretary of the University Committee of the League of Communists of Belgrade; Professor, Faculty of Philology
Dr Yuji Mori, Associate Professor, Institute for Peace Science, Hiroshima University, Hiroshima, Japan
Dr Tetsuro Nakaoka (Professor, Osaka Metropolitan University), Cambridge, England
Dr Vlastimir Novakovic, Dean, Faculty of Mechanical Engineering, University of Belgrade
Dr Cuthbert K. Omari, Associate Professor and Head, Department of Sociology, University of Dar es Salaam, Dar es Salaam, Tanzania
Dr A. N. Pandeya, Professor of Sociology, Department of Humanities and Social Sciences, Indian Institute of Technology, New Delhi, India
Dr Vukasin Pavlović, Professor, Faculty of Political Sciences, University of Belgrade
Dr Miroslav Pečujlić, Rector, University of Belgrade; Chairman, International Seminar
Dr Zlatibor Petrović, Member, Serbian Academy of Sciences and Arts; Professor, Faculty of Veterinary Medicine, University of Belgrade
Mr Dragoman Radojčić, Secretary for National Self-Defence of SR of Serbia
Dr Rasheeduddin Khan, MP, Chairman, Centre for Political Studies, School of Social Sciences, Jawaharlal Nehru University, New Delhi, India
Dr Bruno Ribes, Paris, France
Dr Ing. Slobodan Ristić, Director, Administration for International Scientific, Educational, Cultural and Technical Co-operation of SR of Serbia; Professor, Faculty of Economics, University of Belgrade
Dr Luiz Pinguelli Rosa, Co-ordinator of Post-graduate Engineering Programmes, University of Rio de Janeiro, Rio de Janeiro, Brazil
Dr Pavle Savić, Member, Honorary Board of Seminar; Chairman, Serbian Academy of Science and Arts, Belgrade, Yugoslavia
Dr Vladimir Štambuk, Professor, Faculty of Political Sciences, University of Belgrade
Dr Maksim Todorovic, Vice-Rector, University of Belgrade
Dr Rajko Tomović, Professor, Faculty of Electrotechnical Engineering, University of Belgrade
Dr Kazuko Tsurumi, Professor of Sociology, Institute of International Relations for Advanced Studies on Peace and Development in Asia, Sophia University, Tokyo, Japan

Dr Slobodan Unkovic, Vice-Rector, University of Belgrade
Dr Zoran Vidaković, Professor, Faculty of Political Sciences, University of Sarajevo, Sarajevo, Yugoslavia
Dr Anton Vratusa, Member, Honorary Board of Seminar; Professor, University of Ljubljana; Chairman, Executive Council of SR of Slovenia, Ljubljana, Yugoslavia
Dr Immanuel Wallerstein, Director, Fernand Braudel Centre for the Study of Economies, Historical Systems and Civilizations, Department of Sociology, S.U.N.Y., Binghamton, NY 13901, USA
Dr Ines Wesley-Tanaskovic, Chairman, UN University Council; Professor, Military Academy of Medicine; Member, Committee of National UNESCO Commission
Dr Gazmend Zajmi, Rector, University of Pristina, Pristina, Yugoslavia

The United Nations University

Dr Kinhide Mushakoji, Vice-Rector, Human and Social Development Program
Dr Alexander Kwapong, Vice-Rector, Planning and Development
Dr Anouar Abdel-Malek, Project Co-ordinator, Socio-cultural Development Alternatives in a Changing World (SCA) Project; Co-chairman, International Seminar, Paris, France
Dr Hossam Issa, Program Officer, Human and Social Development Program; Secretary, International Seminar
Mr R. N. Malik, Chief, Conference and General Services; Co-secretary, International Seminar
Mrs Christine Colpin, Assistant to SCA Project Co-ordinator

Name Index

Abdel-Malek, Anouar 40, 58-61, 87, 118-9, 122, 150
*Amin, Samir 42

Barel, Yves 4-8, 21-5, 35-6, 48, 65
Besarović, Vesna 40, 54-8, 61
Blue, Gregory 114, 120-1, 125-30, 146
Bonfil-Batalla, Guillermo 45, 120-2, 136-40, 146

del Campo 146-7
*Ceaucescu, Nicolai 89
*Chomsky, Noam 38

Damjanović, Zvonimir 120, 122-5, 148-9
*Darwin, Charles 74-8
Despić, André 40-1, 60-1, 92
*Dirac, Paul 17

*Einstein, Albert 16-7
*Emmanuel, A. 42

*Frank, A. G. 42
Furtado, Celso 32-3, 68-9, 83, 90, 112-3, 117, 122, 150-1

*Galbraith, J. K. 105
*Gandhi, Indira 64-5

Hassan, Ahmad Yousef 41, 62-3, 114, 126
*Hegel, G. W. F. 11, 82
*Hermann, E. S. 38
*Hiroshige 134
Holland, Stuart 9, 33-4, 36, 61, 69, 84-5

*Imanishi, Kinji 67-8, 74, 76-8, 85
Issa, Hossam 40, 61-2, 90, 117-8

Kawano, Kenji 25, 41, 46-9
el-Kholy, Osama 40, 59, 87-9, 91-7
Kovačević, Zivorad 1
Kwapong, A. 148

Le Thành Khôi 8, 37-8, 117-8
Lefebvre, Henri 8, 10-5, 27, 36, 43, 87, 113
Leite Lopes, J. 8, 10, 15-21, 150
*Lorenz, Konrad 75-6

Macura, Miloš 9, 31-2, 34, 36-7, 65
Maraj, James 40, 59-60, 115
Marton, Imré 147-8
*Marx, Karl 11, 26, 33, 124, 141
Milanović 68, 79-83
*Mohî al-Dîn Câber 5
*Monod, Jacques 71
Mori, Yuji 67-8, 74-9, 81-2, 87
*Motoro, Kimura 71
Mushakoji, Kinhide 38, 111-2

Nakaoka, Tetsuro 41, 121-2, 130-5, 147
*Needham, Joseph 120-1, 125-7, 129

*Ohashi, Shuji 132-3

Pandeya, A. N. 8-10, 25-7, 34-6, 65, 68, 81-3, 85, 88, 114-5, 121, 139, 149
*Pascalon, Pierre 6
Pečujlić, M. 9-10, 35-6, 120, 122, 140-6
Pinguelli Rosa, Luiz 2, 20, 88, 90, 107-11, 120, 149-50

Rakić, L. 68, 85-6
Rasheeduddin Khan 63-5, 90, 115-7
*Rezvani 5

Ribes, Bruno 67, 69-74
Ristić, Slobodan 39-40, 49-53

*Said Al-Andalusi 62-3
*Sakuma, Shozan 46
Savic, Pavle 2-3
Silva-Michelena, José 40, 88-9, 97-101
Štambuk, Vladimir 9, 36, 39-46, 59, 61, 87, 113-4

*Tokugawa, Ieyasu 148
Tomović, Rajko 8-9, 24, 27-31

Tsurumi, Kazuko 68, 85, 131

*U Thant 32

*Vessuri, H. 100
Vidaković, Zoran 68, 88-90, 101-7, 120, 122, 140-6

Wallerstein, Immanuel 39, 65-6, 69, 146
*Weber, Max 149
*Wilson, Edward 75

Note: This index includes participants at the conference and also other personal names mentioned in the book. The latter are marked with an asterisk*.

Subject Index

Algeria, traditional peasantry 5
Algorithmisation 24-5
Angola 99-100
Arms Race, *see* Military Spending

Bio-Sociology 76-7, 81-2, 85
Biology 24, 67-8
  knowledge of
    and control over life 69-71, 73-4
    need for 69-70
  research into
    bacteriological 70
    behavioural 71
    and definition of life 71-4
    direction of 67, 70-71
    genetic 70-71
    guidelines for utilisation 67, 73-4
  role of 35
  theory
    nature of 67
    and social theory 74-5
Brazil 118
  hydroelectric power 108
  nuclear energy in
    future construction programme 108
    and German technology 109-10
    and Latin-American cooperation 110
    national control of 109-10
    need for 108-9
    and nuclear weapons 110-11
    problems concerning 109

Capitalism 18-9, 29, 33, 36, 41, 97-8, 112-3, 142
  developmentalist ideology of 45-6, 99-100
China 32, 59, 62, 97, 99-100
  traditional science
    and bureaucratic feudalism 127-8
    compared with Europe 126
    factors facilitating 129
    incorporation into modern science 127-30
    medical knowledge 129-30
    theoretical contributions 126-7
  traditional technology
    factors facilitating 129
    and transmission to the West 126
Citizen, concept of 14-5
Colonialism 37, 54-5, 60, 64, 117-8, 129, 137, 141
Consumer, concept of 14-5
Creativity
  endogenous 1, 58, 65
  importance of 125
  reawakening, in Third World countries 114-5

Darwinism 74-6
Decentralisation 15, 25, 38, 49, 111
Democracy 2, 15, 22, 27, 29, 58, 94, 125, 146
Developing countries
  and appropriate technology 19, 34-5, 65
  collective power of 115
  dependency of 49, 102-3
  and developed countries, gap between 105, 125
  development of, technological inhibition 102
  endogenous creativity in 65
  energy policy for 107-8
  higher education in 52-3
  historical classification of 62-3
  impotence of 121
  inventions from 54
  nuclear energy in 109

Developing countries (*cont.*)
  pooling of resources in 19-20
  research and development activities in 51-2
  scientific-technological development in
    and collective self-reliance 39-40, 50-53
    and dependence on developed countries 49
    financing of 53
    leaps in 133-4
    problems of 52
    strategies for 50-1
    and technology transfer 49-50, 53
  self-reliance of 39-40, 50-3, 65-6
  social movements in 98-9
  technology transfer to
    compensation for 56-7
    control of conditions for 55
    legal regulation of 53, 55-7
    negative effects of 4, 49-50, 54-5
  and transnational corporations, direct dealing with 101
  and transnationalisation, consequences for 98
  and UN development system 53
Development
  conceptions of 39, 41-3, 61, 100, 114-5
  global inequalities in 63-4
  historical dimension 148
  and industrialisation 63, 133-5
  role of science and technology in 1, 44-6, 92-7, 122-3, 133-5
  see also Third World countries

Education 28-9, 124-5, 148
  effect on labour force 147-8
  and manipulation 29
  role in loss of ethnic identity 138-9
Energy
  domestic consumption 108
  nuclear, *see* Nuclear energy
  policy for developing countries 107-8
Eritrea 89, 100
Eurocentrism 11, 37, 43, 87, 149
Europe
  early science in, compared with other cultures 16, 126
  exogenous elements in development of 131

  modern science in
    and Chinese traditional science 129
    domination 121, 128-9
    social background in development of 146-7
    synthesis of 127
    technological development in, compared with Japan 132-3
Evolution
  Darwinian approach 74-5
  debates over 71-2
  non-Darwinian theory of 76-8, 85
  and over-population 77-8
  types of 72
  vertical 72-3

Genetic control 70-71, 73
Germany 18, 33, 54, 100, 109-10
Globalisation 8, 10-13, 33, 39, 41, 64-66, 94, 113, 126, 148
Great Britain 33

Health 35-6
Health care 30-1, 85-6, 145
  need for a multidisciplinary approach to 68, 86
  and transnational corporations 68, 82-3, 85
Hegemony 58, 102-6, 141, 145-6
  and the arms race 90
  definition 87
  role of science and technology in 88, 102
  struggle against 87-9
Hereditary diseases 69
Human behaviour
  aggressive 75-6
  research into 71

Imperialism 117-8
  see also Hegemony; Transnational Corporations
India 62, 118
  black market economy 116
  contributions to modern science 127
  cultural minorities in 149
  scientific creativity of 58

Subject Index

Industrialised countries
  and developing countries, gap between 105, 125
  employment problems 9, 33
  labour movements in 112
  problems facing 9, 61
Information systems
  control of, by the state 13-5
  and political power 13-4, 37-8, 105, 118
Information technologies 30, 38, 64, 111-2
Intermediate technology, see Technology, appropriate
Iran 89, 117
Island communities 59-60
Italy 35-6

Japan
  government
    centralisation 46
    decentralisation 48-9
    Meiji 46-7, 131-2
    role in scientific-technological development 47-9
    support of universities 46-7
  science and technology in 54
    government role in development 47-9
    negative effects 47-8
  technological development in
    exogenous character of 131
    leaps in 133-5
    and self-reliance 134-5
    and social changes 134-5
    uneven nature of 135
    views of 130-1
  technology, imported
    conflicts resulting 131-2
    and indigenous technology 133
    and speed of development 132-3
  US influence in 47
  universities 46-7
  views of science 130-1, 134

Knowledge industry 34
Korea 89, 100

Latin America
  early science in 17
  economic system of 17-8
  endogenous development 139-40
  Indian groups in
    ethnic identity, reaffirmation of 138-9
    loss of ethnic identity 138-9
    natural knowledge of, 17
    social knowledge of 137-40
  multinational corporations in 19
  nuclear programme, need for 110
  social movements in, control of 99
  technology transfer 18-9
  traditional and modern science in 130
  universities 18

Mass production 28-9
Mathematics
  experimental 124
  and scientific method 128
Medicine
  bureaucracy of 80
  Chinese 129-30
  individualistic approach to 35, 67
  place in social life 81
  specialisation in 80
  transnational corporations, role in 68, 82-3, 85
  see also Physicians
Mental diseases 69-70
Mexico 138-9
Military spending 39, 51, 64, 68, 83, 85, 88-90, 94, 101-107, 141
Minamata 85, 135
Multinational corporations, see Transnational corporations

Natural resources, squandering of 9, 19, 29, 32, 64, 86, 107
Non-aligned movement 58, 89-90, 101
Nuclear energy
  in Brazil 108-11, 112
  and nuclear weapons 90, 110-11
  problems, in developing countries 88, 109
Nuclear weapons, proliferation of 110-11

Oecumenical science, see Science, universality of
OPEC 99, 100, 117
Over-population 77-8

Physicians
  responsibilities
    complexity of 79
    concept of 68, 81
    and professionalism 80-1
    and social expectations 79-80
    and specialisation 80
  see also Medicine
Physics 20, 24
  development of 16-7
Post-industrial society 13
Power
  basis for 117-8
  changes in, worldwide 112-3, 118
Profit optimalisation 29, 32, 35

Research and development 124
  in developing countries 51-2, 95

Science
  as cultural force 6, 26-7
  as force for transformation 114-5
  forms of 26
  as liberating force 20-1, 26-7
  modern
    definition 128
    differences from traditional science 128-9
    European domination of 61, 121, 128-9
    relationship with traditional science in developing countries 130
    synthesis of 127
  nature of 20, 123
  portrayal of 7-8
  see also Scientism
  traditional 62-63
    differences from modern science 128-9
    and theory 126-7
  universality of 120-2, 124-5, 128-30, 145-6, 150-1
  views of 91, 130-1, 134
Science and technology
  benefits of, 2-3
  contradictory aspects of 1-2, 141-2
  control of, by superpowers 2
  as cultural components 38, 120, 124
  definitions of 43, 128
  as democratic factors 125
  development of 38, 88
  aims of 32
  creativity in 58-9
  financing of 52-3, 94, 151
  in Japan 46-9, 130-1
  knowledge base needed for 149
  nations classified according to 62-3
  options for, in Third World countries 91-2, 116
  orientation of 103-4
  and political power 58, 146, 150
  as prerequisite for economic independence 122-5
  problems in 52
  and self-reliance for developing countries 39-40, 50-3, 121-2
  social determination of 7-8, 59
  and social structure 63, 146-7
  strategies for 44-5, 50-1
  and transfer of technology 49-50, 53, 93
  and world market, effect on 8, 12
  distinction between 59, 149-50
  European particularism and 78, 92-94, 114, 116, 121, 124, 126-7, 128-30, 131, 146-7, 149-51
  in Japan, historical development of 46-9
  as liberating force 10
  and emancipation of the labour force 143
  and living conditions 145
  and modes of production 122, 144-5
  potentials of 141
  success of 145-6
  and transformation of labour processes 143-4
  and unemployment 144
  monopolistic control of 49, 51, 141-2
  purpose of 103-4
  and repression 51, 101-2, 105-7, 141
  role of leading corporations in 105-7
  social components of 2, 105-7
  objectivity of 120-1
  proficiency in, as alleged European trait 43, 59-60, 62-3, 121, 124
  relationship between 27-8, 38, 59
  relationship with society 21-5, 123-5, 142-3

Subject Index

Science and technology (*cont.*)
  repressive nature of
    and dependence of developing countries 102-3
    effectiveness 102
    expansion of 103-4
    and exploitation of the labour force 104-5
    and militarisation 51, 68, 94, 101-4
    structure of 101
    tendencies toward 103
    transformation of 103
    use of, by ruling classes 101-2
  role 21
    ambiguous nature of 2
    in development, views of 44-6
    as formative factors 4-7
    in human liberation 141-6
    in social change 91, 93-4, 96-7, 143
    and self-determination 21-5, 92-6
    and social development 8-9, 78-9
    social functions, mystification of 101
    and social identity 6-7
    universality of 124
    view of 94
  see also Science; Technology
Scientism 7-8, 13, 88, 94, 121-2, 150
Self-determination 15, 60
  effect of science and technology on 21-5, 36
  and mechanisation 24-5
Self-reliance, national
  aim of 39, 50, 93-4
  as approach to problems of development 45, 93-4
  in developing countries 9
    collective 19, 39-40, 50-3
    importance of 50, 88
    problems 65-6
Social identity, loss of 6-7
Social knowledge
  and classification of people 136-7
  institutionalisation of 140
  in Latin American Indian societies 137-40
  role in social development 136, 140
  and social science 137, 140
Social passivity, and modern technology 5-7
Social sciences 26-7, 94, 120, 146-7
  conception of 136
  endogenous participation in 26-7, 94, 136
  relation to biological sciences 74-5, 86
  role in colonial systems 120, 138
  and social knowledge 137, 140
  universality of 138, 146-7
Socialism 2, 27, 42, 45-6, 95, 142
  world transformation toward
    and economic transformation 89, 97-9
    and political transformation 89, 99-100
    and transnationalisation 98
Socio-biology 75-6
Spain 147
Students
  and intellectual creativity 147
  psychological control of 71
  in Third World countries 4-5

Technology
  appropriate, *see* Technology, appropriate
  areas of demand 30-1
  and attitude to health 35-6
  characteristics of 32
  development of
    basic social goals 28-9
    financing of 32
    historical 28-9
    in Japan 131-5
    and military spending 68, 83
    need for leaps in 133-4
    negative consequences 84
    prospects for 29-31
    and protectionism 84-5
    and social control 84-5
  and employment problems in industrialised countries 9, 33
  exchange of 93
  as feature of Western civilisation 43
  as generator of power 30, 83
  and national creativity 1
  potentials of 30, 83
  role in dependence of developing countries 4, 18-9, 100
  and social passivity 5-7
  and social stratification 77
  traditional Chinese 126, 129
  transfer 40, 61, 93, 100-1
  and conflict with indigenous culture 131-2

Technology transfer (*cont.*)
  control of, by suppliers 19, 55
  and dependence of developing countries 4, 18-9, 54-5, 100
  influence on local population 5-6
  to Japan 132-3
  to Latin American countries 18-9
  legal regulation of 53, 55-7, 61-2
  negative effects 49-50, 54-5
  and unemployment 69, 84
  see also Science and technology

Technology, appropriate 9-10, 19, 45, 58, 63, 65, 109, 120
  development of 32-3
  and national dependence 9, 34-7, 100
  need for in developing countries 9, 19, 31-32

Third World countries
  control of power resources in 85, 88, 112, 117
  cultural domination of 37-8, 92-3
  divisions within 90-1
  domination by USA 89
  economic potential 85-6, 90
  education in 37
  political differences 116-8
  and relationship with USSR 89
  scientific-technological development in
    and creativity 114-5
    factors in 121-2
    knowledge base needed for 149
    options open to government 116
    questions arising from 116-7
    see also Appropriate Technology
  self-reliance of
    benefits 93-4
    need for 88
  socio-economic systems, role of 92
  transformation of
    global perspective of 91-2
    and national inheritance 92
    political prospects for 95
    role of science and technology in 91, 93-4, 96-7
    role of socio-political systems in 92
    and self-reliance 93-4
    stages in 96-7
    and Western culture 92-3
    see also Developing countries; and under specific countries

Transnational corporations 8, 39, 62, 64, 88, 95, 101, 113, 115, 150
  character of 105-6
  and control of national economies 18-20, 100-101, 113, 115-6
  extent of power 106
  in Latin America 18-9
  and metropolitan technocracy, growth of 105
  relationship with the military 106-7
  resistance to 113-4
  role in globalisation 64
  role in health care 68, 82-3
Transnationalisation 66, 98, 113
  see also Globalisation
Treaty of Tlatelolco 111

Underdevelopment 42
Unemployment 31, 33, 144
  and technological innovation 69, 84, 124
United Nations 53, 116
United Nations Conference on Science and Technology for Development (UNCSTD) 32, 91, 148
Universities
  in Japan 46-7
  in Latin America 18
  and social needs 1-2
USA 34, 47, 51, 54, 64, 89
  confrontation with USSR 89, 97-100
  and nuclear fuel, control of 110
USSR 51, 54
  logistic support of revolutionary movements 89, 99-100
  and USA, confrontation with 8-12, 89, 97-100

Western civilisation 43, 92-3
World Bank 34, 64, 116
World-views 8-12, 15-16, 23-5, 113-4
World economy
  militarisation of 88-90
  trends in 83
World Market
  analysis of 12, 83, 112-3
  effect on science and technology 8, 12, 126

Yugoslavia 36, 89, 123

Zimbabwe 89-90